GLOBAL ENERGY INNOVATION

GLOBAL ENERGY INNOVATION

Why America Must Lead

Woodrow W. Clark II
and Grant Cooke

AN IMPRINT OF ABC-CLIO, LLC
Santa Barbara, California • Denver, Colorado • Oxford, England

Library of Congress Cataloging-in-Publication Data

Clark, Woodrow W.
Global energy innovation : why America must lead / Woodrow W. Clark II and Grant Cooke.
p. cm.
Includes bibliographical references and index.
ISBN 978-0-313-39721-9 (hardback) — ISBN 978-0-313-39722-6 (ebook)
1. Power resources—Research—United States. 2. Power resources—Research—International cooperation. 3. Energy policy—United States.
I. Cooke, Grant, 1947- II. Title.
TJ163.25.U6C585 2012
333.79'4—dc23 2011030730

ISBN: 978-0-313-39721-9
EISBN: 978-0-313-39722-6

16 15 14 13 12 1 2 3 4 5

This book is also available on the World Wide Web as an eBook.
Visit www.abc-clio.com for details.

Praeger
An Imprint of ABC-CLIO, LLC

ABC-CLIO, LLC
130 Cremona Drive, P.O. Box 1911
Santa Barbara, California 93116-1911

This book is printed on acid-free paper ∞

Manufactured in the United States of America

This book is dedicated to Woody and Grant's families, children, and grandchildren, with more expected. Since they are likely to benefit from it the most, *Global Energy Innovation: Why America Must Lead* is dedicated to them.

Contents

Acknowledgments ix

Introduction 1

1. A New Carbonless and Sustainable Era Is at America's Doorstep 7
2. The Problem: Climate Change and Its Global Impact 23
3. Asia Leads the GIR: Japan and South Korea 41
4. Europe Joins the GIR 53
5. China Leapfrogs into the GIR 63
6. Sustainable Communities 77
7. Renewable Energy Integrated Systems 97
8. Smart Green Grids in the GIR 117
9. Emerging Commercial Technologies Empower the GIR 129
10. The Next Economics 149
11. The Race for Global Energy Innovation Has Begun: Where Is America? 167

References 179

Index 189

Acknowledgments

Above all, our families are the first to thank for their patience and support of this book.

Woody's wife, Andrea, was and is a strong supporter of this book from the beginning. However, Woody III, Woody's older son by his first marriage, was the catalyst for doing this book. He heard Woody II give a talk on the Green Industrial Revolution (GIR) in the fall of 2009. While Woody III had heard his dad talk and give speeches his entire life, his comment after this talk was, "Dad, now I understand what you have been talking about all these years. This Green Industrial Revolution makes sense." Then he added, "Dad, you need to do a book on this topic." The result: this book on global energy innovations, with a strong focus on the GIR.

It's also important to note that the origins of the idea about a third industrial revolution came from a series of workshops that Woody worked on with his friend and colleague Jeremy Rifkin for almost a decade. Rifkin, president of the Foundation on Economic Trends, first used the term the "Third Industrial Revolution" in his book *The European Dream* (2004), but did not want to do a book exclusively on the topic. However, in the spring of 2010, Woody wanted to do a book on the topic and approached Grant about co-writing it because they had each been working on consulting projects in the sustainable communities area.

The world's response to climate change will have an unprecedented impact on the course of human history. The shift to mitigate global warming and equally disastrous environmental changes will create unimaginable social, economic, and political change—just as the First and Second Industrial Revolutions did. With almost seven billion people now on this fragile planet, the sands of time are marking our existence if we continue to use the environment and atmosphere as garbage cans. The GIR carbonless economy, driven by nonpolluting renewable energy technologies, sustainable communities, and resource conservation, offers a rare chance to reverse the environmental degradation caused by human dysfunction and mismanaged resources.

While intelligence, experience, and insight can help in the prescient recognition of a megatrend like the GIR, hard work is what captures the transformative ideas and hones them into a coherent thesis.

This hard work is impossible without the support of family, and Grant would like to thank his wife, Susan, for her forbearance, patience, and insight as this manuscript came together. Grant has children, Brian and Lauren, and grandchildren, Ayla and Revely, who he hopes will benefit from a healthier world not ravaged by global warming or carbon-based pollution.

Woody and Grant also thank Valentina Tursini, their acquisitions editor at Praeger. She liked the idea of this book from the beginning and made it a reality. She has been both an incredible supporter and constructive critic. Her dedication and critical insights have helped make this book into what the authors believe will be a classic point of reference for future generations. With her background in philosophy, Valentina is sensitive to today's big themes and issues and is interested in acquiring books that will have an impact on our society. She was immediately able to recognize the importance of this project, as it called for a global change in the way we interact with our environment. She has been a great supporter from beginning to end.

Woody and Grant would also like to thank Nancy Sakaduski, their developmental editor, who played a key role in helping them clarify their message and make complex information accessible to a broad range of readers. Her insights, changes, and additions were essential to putting the book into final form. Nancy is herself the author of more than 100 articles and 15 books (most under the name "Nancy Day") and has just published the third edition of *Scientific English* (as co-author). Her suggestions and assistance were greatly appreciated.

Finally, the authors wish to thank the many people behind the scenes. One group stands out in particular: the creative marketing staff at Praeger, who designed the cover for the book. The authors wished to emphasize more than just the common everyday symbols of renewable energy—solar panels and wind turbines—believing that global energy innovations are the impetus and evidence that a GIR already exists. So the cover needed to be much more.

Beyond solar and wind, the authors have added advanced technologies from Horizon Fuel Cells in Singapore and China, along with a UK new hydrogen fuel-cell car called Riversimple, which was newly released in 2009. What is significant about the photo of the Riversimple car, however, is the people around it. For the authors, that photo symbolizes what this book is about: The world as it can be today and not in 20–50 years. In fact, the world must be like these photos if we are to survive and reverse climate change. Our children's and grandchildren's lives depend on our acting now. Hence the book's subtitle: *Why America Must Lead*. Now.

Introduction

The world is entering an extraordinary new era that will result in a complete restructuring of the way energy is generated, supplied, and used. Called the Green Industrial Revolution (GIR), this will become the largest social and economic change of the postmodern era. It will be an era of extraordinary potential and opportunity, with remarkable innovation in science and energy that will lead to sustainable and carbonless economies powered by advanced technologies like hydrogen fuel cells and nonpolluting technologies like wind and solar. Small community-based renewable energy generation will replace massive coal and nuclear-powered utilities, and smart grids will deliver energy effortlessly and efficiently to intelligent appliances. America must shake off the angst of 9/11 and become the world leader in global energy innovation to reinvigorate its dormant entrepreneurial spirit and stimulate its sagging economy.

This new social-political-economic era encompasses changes in technology, economics, businesses, jobs, and consumer lifestyles, and will be a fundamental change from the Second Industrial Revolution (2IR), which was based on fossil fuels, internal combustion engines, mechanical engineering, and neoclassical economics. This monumental shift is already underway in Europe and Asia, where renewable energy, sustainable communities, smart green grids, and environmentally sound technologies are becoming common. Now, it is at America's doorstep.

The concept of the Third Industrial Revolution was referred to by Jeremy Rifkin in his book *The European Dream* (2004), where he noted that the development of the modern world was marked by extraordinary changes brought about by the nexus of digital communications and renewable energy to power buildings and vehicles. Rifkin identified a Third Industrial Revolution in Europe during the 1990s. His book was devoted to acknowledging European leadership in business, economics, and environmental development at the end of the 20th century.

Although Rifkin saw this change only in Europe, the beginnings actually developed earlier, in Japan and South Korea. In the 1980s, these Asian governments were concerned with the need to become energy secure, and, as a result, they developed national policies and programs to reduce their growing dependency on foreign fuels. These countries realized after World War II and the Cold War that their futures were not rooted in the same carbon-intensive economies that had built the United States and Western Europe. These Asian nations were too highly dependent on imported fuels and energy to grow rapidly. In fact, the Russia–Japan war of 1905 and both World Wars were rooted in the need for Japan to be energy independent.

A century later, Japan is once again struggling with an energy crisis, created by a devastating earthquake and tsunami in its northeast coastal region that destroyed one of its key nuclear power plants in Fukushima. From this tragedy, Japan may leap even further ahead in developing a carbonless economy as it expands its renewable energy generation to compensate for the loss of nuclear power.

Despite all the activity in Europe and Asia, few Americans (outside of a small circle of scholars and a handful of prescient venture capitalists and investment bankers) saw this new global megatrend looming until recently. Even many people within the green industry have remained oblivious, such as engineers and economists who are famous for their tunnel vision. For the most part, Americans failed to see that the carbon-based Second Industrial Revolution was ending, because they had become used to the dependable energy and comfortable lifestyles it had supplied for so long.

The goal of this book is to bring this new era with its remarkable opportunities and extraordinary benefits into the light of America's day to point the way to America's future. A change to a new carbonless economy will require a massive transfer of resources, away from a lifestyle powered by fossil fuels and coal-based energy sources and toward one that relies on renewable energy distributed by smart grids. This renewable energy will be community-based, flexible, and environmentally friendly, although it will probably be more expensive in the near term. The primary drivers for this transformation are the fact that: (1) the world's oil supply and other cheap carbon-based fuels are in decline; (2) climate change and global warming are a major environmental threat; (3) the emerging technologies for renewable energy power generation are dramatically improving efficiency and cost-effectiveness; and (4) America badly needs the new "green" jobs that will be created by this era of global energy innovation.

This transition to carbonless energy generation will be a challenge to America's "old" economy, but will help create a new green economy. Previous ways of creating wealth for America's businesses and shareholders were natural resource extraction, fossil fuel–based power generation, and energy-intensive manufacturing. As a corollary, economic growth was rooted firmly in land development. As Clark and Fast discussed in *Qualitative Economics: Toward a Science of Economics* (2008), such neoclassical economics are rooted in simple models of supply and demand, which had to balance in the short term due to market forces rather than to government involvement, oversight, or regulations. These cost-benefit analyses were directly linked to corporate earnings, executive bonuses, and financial incentives.

The businesses that generated the bulk of America's wealth—from the first colonies until the 1990s—were dubbed the "dirty rich," while knowledge and service companies that generate little pollution have been called the "clean rich." After a long reign at the top of the economic ladder, the dirty rich are finally being displaced. David Callahan, author of *Fortunes of Change: The Rise of the Liberal Rich and the Remaking of America* (2010), points out that of the top 20 companies on the Fortune 500 list in 1960, 16 were engaged in heavy industry (companies like U.S. Steel and DuPont), or resource extraction (companies like Texaco and Mobil). By 2009, only six such companies were in the top 20 (Callahan 2010).

The number of individuals who are getting rich from fossil fuels is also dropping—even in Texas, the historical center where fossil-fuel fortunes were made. Fewer than half of the state's current billionaires on the Forbes 400 list of the wealthiest individuals made their money in oil or energy, which is a major departure from the past (Callahan 2010). Two of the wealthiest Texans are not oil magnates: one is Alice Walton, heir to the Wal-Mart fortune, and the other is Michael Dell, founder of Dell Computers. Even more significant, it is now Texas—not California—that leads the nation in installing and generating wind power, one of the key components of renewable energy of the new era.

The condition of the American economy is also playing a role in driving this green industrial revolution, which offers the promise of new jobs—most likely "green" jobs. This could provide relief to an economy that has been stagnating with chronically high unemployment.

The era of sustainability and renewable energy has begun. America's transition, from an economy based on the extraction of natural resources and fossil fuels to one based on knowledge and renewable energy, is gaining momentum. It is clear that there are multiple drivers for this looming megatrend.

Industrial Revolutions Mark Historical Social Change

The First Industrial Revolution was a turning point in human history. Most historians say that it started in England around 1760 and ran through the last part of the 19th century. Great Britain led the world as it transitioned from a manual labor, agricultural–based economy to machine-based manufacturing. There were major changes in agriculture, manufacturing, mining, warfare, transportation, philosophy, and especially communications.

New and startling ideas drove the Age of Enlightenment, and the printing press made it possible to communicate them. Also providing thrust for this new industrial society was the Watt steam engine. James Watt and his business partner Matthew Boulton dramatically improved a 1712 design, and popularized the coal-burning steam engine, which ushered in the great textile and manufacturing industries. Soon, all of Western civilization was entrenched in the First Industrial Revolution, driven by steam, and surrounded by innovative ideas distributed with the help of the printing press.

The internal combustion engine powered by fossil fuel was at the heart of the Second Industrial Revolution. Although various forms of the internal combustion engine were developed before the 19th century, it only became a practical invention with the start of commercial oil drilling and production in the mid-1850s.

Along with machines, electricity, and transportation came a technology that revolutionized the daily lives of ordinary people: the telephone. There is some controversy surrounding its invention, including a claim that an Italian, Antonio Meucci, was the actual inventor, rather than Alexander Graham Bell. However, it was Bell who spoke the first complete sentence, which was transmitted on March 10, 1876. To Thomas Watson, his assistant, Bell said across the line, "Watson, come here; I want you." The commercialization of the telephone was an iconic example of American entrepreneurship.

The Third, or Green, Industrial Revolution is more significant and life changing than either the First or Second Industrial Revolutions. There is so much more at stake. Today, the world is rapidly running out of fossil fuel. This alone threatens to shake the very foundation of human existence. Adding a heightened sense of urgency is the environmental degradation and the collapse of various parts of our planet's ecosystem.

Climate change is real. Almost all of the world's scientists have accepted the work of the United Nations Intergovernmental Panel of Climate Change (UN IPCC) that was the co-recipient of the 2007 Nobel Peace Prize. The UN IPCC did groundbreaking work that documented that humanity was responsible for global warming. The four UN IPCC Reports have now been accepted by the world's biggest fossil-fuel dependent nations, including America.

Today, the planet is being threatened by melting ice caps, ocean acidification, and extreme weather patterns, and is losing sensitive and critical environments. As part of this environmental degradation, numerous animal species are lost each year. The International Union for Conservation of Nature (ICUN) estimates that 40 percent of the planet's known organisms are considered endangered. In 2009, the ICUN reported that 21 percent of all known mammals, 30 percent of all known amphibians, 12 percent of all known birds, 28 percent of reptiles, 37 percent of freshwater fishes, 70 percent of plants, and 35 percent of invertebrates assessed so far are under threat. From Uganda's mountain gorillas to California's native trout, the world's animal species are threatened by the detritus of the era of cheap energy and unsustainable lifestyles. Humans simply must stop using the environment as a massive garbage can.

The premise of this book is that it is not too late to reverse this trend. It will require changes in attitudes, habits, technologies, business practices, government legislation, and economics, but these changes are already being made in many parts of the world. Their successes prove that these new approaches work and can be integrated successfully. America must not only commit to this effort, but also become a leader. A world of renewable energy and sustainable communities is the only hope for the future.

Chapter 1

A New Carbonless and Sustainable Era Is at America's Doorstep

Americans like to think of their country as the world's technological leader. After all, America led the Second Industrial Revolution (2IR), the technological transformation that produced steel to build skyscrapers, railroads to carry goods across the country, electricity to light up cities, and chemicals to make daily life easier. By the end of the 20th century, America was the world leader in innovation and entrepreneurship, creating historic advances in computerization and information technology. Now, that distinction as an innovator and entrepreneurial dynamo is gone. When the world sought leadership in the battle to stop global warming and reverse climate change, America sat on the sidelines.

The Green Industrial Revolution (GIR)—one that is creating the technologies for the renewable communities of the future—has begun. What hasn't changed is America's reliance on 2IR energy sources, business models, and technologies. Environmental economist Jeremy Rifkin came up with the concept of the Third Industrial Revolution in his book *The European Dream* (2004) as first coming from Europe in the 1990s. However, it actually began in Japan and South Korea many decades earlier and has leapfrogged into China. The Green Industrial Revolution has yet to come to America.

America, as a nation and as the leader of democracy for two centuries, must examine its own roots and provide the future direction for humanity. U.S. leadership must reestablish global vision with strategies that move it toward a carbonless economy and away from its oil and gas addiction.

The GIR is arriving as the Internet, with its social networks and digital electronics, intersects with renewable-source energy distributed by smart grids. Social and economic forces are coming together as the United

States ponders its environmentally sustainable future. Now with global warming and climate change impacting daily lives, can the nation wait any longer?

Asia and Europe have been moving toward it for two decades, developing sustainable, energy-independent communities. Denmark has cities that by 2015 will be energy independent with renewable energy power and smart green grids. South Korea has urban regions that are already energy independent and carbon neutral (Kim 2010).

To remain competitive, America must reject the heavy-carbon cloak of the 2IR, with its massive and inefficient fossil fuel generation, and move rapidly into the GIR, with its community-centric and environmentally friendly renewable energy generation. As a first step, America needs to reduce its energy dependency on the Mideast, a geopolitical region with such instability that it constantly threatens national security, drains precious financial resources, keeps the nation from focusing on crucial domestic issues, and takes thousands of lives.

The situation is complicated by the continuing 2008 economic crisis. The world is battling the most severe economic turndown since the Great Depression of the 1930s. The turmoil in the Middle East could move the world into an even a deeper recession. American states are reeling with the loss of tax and real estate development revenue. Yet, the basis for the economic downturn is rooted in the older neoclassical economy. While Europe and Asia see that fossil fuels are not their economic future, America continues to invest in offshore drilling, shale oil, nuclear power, and massive transmission systems.

In 2011 California was ranked as the world's eighth largest economy. Yet before deregulation of energy became law in 1996, according to the U.S. Department of Commerce, the state was the world's sixth largest economy and held the distinction as number seven from 2003–2008. At the turn of the 21st century, California experienced a serious energy crisis that continues a decade later. Now California is on the verge of bankruptcy, just as General Motors was in 2009, primarily because of the economic dependency on 2IR technologies and the fact that business interests that support those technologies continue to get tax breaks.

In 2010 Toyota overcame General Motors as the world's number one carmaker, due in large part to the Toyota Prius, the car that commercialized hybrid technology. The Prius uses a combination of electricity and gasoline to get an average of 45 miles per gallon. It has a regenerative braking system that takes electrons produced from applying the brake and stores them in the cars' batteries. The regenerative braking technology was invented in the

U.S. Department of Energy (DOE) national laboratories in the 1990s. The DOE offered the technology to American carmakers in the mid 1990s, but all of them turned down the right of first refusal. Toyota saw its value and licensed the technology. The rest is history, including the fact that in 2008, Ford licensed the regenerative braking technology back from Toyota for use on its own hybrid vehicles.

The United States has made many extraordinary innovations in environmental technologies that have led other nations into the Green Industrial Revolution. But the technologies were either ignored or replaced by the economic and political forces in America that represent and are invested in technologies from the 2IR.

Too Little Oil, Too Much Demand

Oil is a product of ancient life millions of years ago. It is made deep in the earth, from the crushed remains of rotting organic matter—animal, phytoplankton, plants, and algae. It takes millions of years with just the right temperature and pressure for this material to cook down into oil. Too much sediment on top of the remains will overcook the matter and turn oil into natural gas or unusable material. Too little sediment means no oil at all. Finding oil has always been a wild and exciting explorative process. For more than a hundred years of the 2IR, oil and gas discoveries have made fortunes for those greedy and smart enough to understand the earth's geology. Or, as in the case of the Middle East, lucky enough to live on top of enormous hydrocarbon deposits.

Now, scientists believe that the world's oil and natural gas supplies have peaked and are rapidly declining because of the demands from the technologies of the 2IR. As M. King Hubbert, the Shell Oil geophysicist, observed in his startling prediction first made in 1949, the fossil fuel era will be of very short duration. In 1956 he predicted that U.S. oil production would peak about 1970 and then decline (Hubbert 1956). At the time, people scoffed at him. Now, in the early 21st century, Hubbert looks extraordinarily prescient.

According to United States Geological Survey (USGS) data, the earth originally held between five and six trillion barrels of oil. In 2000 the USGS estimated that all but 650 billion barrels of oil had been found (USGS 2003). Scientists think that most of that undiscovered oil is hidden deep under the sea in basins trapped in salt substructures. In deep water, the salt masks the hydrocarbons, making exploration expensive, tiresome, and exceedingly complicated. It was under these salt layers that the vast Tupi

Field in Brazil was discovered in 2006, and scientists now are probing deep underwater in Angola in similar geology.

Richard Sears, a geophysicist and former vice president for exploration and deepwater technical evaluation at Shell, appeared at the 2010 Technology Entertainment and Design (TED) conference in Long Beach, California. In his talk on the future of energy, Sears said that there are only 30 to 50 years left before a broad gap opens between worldwide oil supply and demand. Sears then held up a pincushion of the globe with red thumbtacks. "This is it," he said. "This is all the oil in the world. Geologists have a pretty good idea of where it is" (Sears 2010).

Throughout the world, oil supplies are in decline (Sears 2010). For example:

- Kuwait's oil supplies have been in decline since 1970.
- U.S. oil supplies have been in decline since 1971, hence its growing dependency for foreign oil and gas, along with the need to protect those resources around the world.
- Iran's oil production has been in decline since 2008, while demand is soaring.
- Indonesia's oil supplies have been in a free fall since 1991, and this former OPEC exporter is now an importer.
- The European North Sea oil reserves have been declining since 1999, and the decline is accelerating. The United Kingdom, once an oil exporter, is now seriously talking about oil rationing.
- Norway's oil production has been in decline since 2001, and for environmental reasons has limited and cut back on its offshore drilling.
- Mexico's oil production has been dropping since 2005.
- Canada has increased its shale oil drilling, while destroying thousands of acres of virgin land.

While oil supply decreases, demand increases. In early 2011, China released customs data that showed that oil imports rose 18 percent in 2010 (Energy Sector Investing 2011). Platts, an oil industry research company, reported that China's oil consumption in March 2011 averaged 9.2 million barrels per day. Platts calculates that China's apparent oil demand is up 10.5 percent year to year. This was much higher than the 8.8 percent predicted by the U.S. DOE Information Administration. A report released around the same time from oil giant British Petroleum said that China will be the largest source of oil consumption growth over the next 20 years. China's consumption is expected to rise to 17.5 million barrels per day, which means that China will overtake the United States as the world's biggest oil consumer (British Petroleum 2011).

Driving this consumption is China's adoption of the automobile. Once a nation where everyone commuted by bicycle, China now has 60 million cars on the road, with 12 to 18 million more new cars predicted for 2011 (British Petroleum 2011). China's immediate solution to the demand problem, while it implements its next five-year plan, has been for its state-controlled oil and gas company to buy massive amounts of oil and gas from around the world on a long-term basis and import them to the country.

India is not far behind, consuming nearly three million barrels per day. Car sales jumped 21 percent in November 2010. India is expected to be the fourth largest car market in the next three years ("India Car Sales Jump 21 Percent" 2010). Meanwhile in both India and China, pollution increases and emissions grow, spreading around the world and causing climate change.

As the world's oil and natural gas supplies peaked, interest in nuclear power grew. This is a false hope, particularly in light of the tragic 2011 earthquake and tsunami that struck and disabled the nuclear power plant in Fukushima, Japan. In 2001 the U.S. DOE provided a key set of figures documenting the declining supplies of gas and oil for 60 and 50 years, respectively. The surprising statistic, and one that bodes ill for the supporters of the nuclear power industry, is that there is less than a 60-year supply of uranium left. Now, a decade later, the DOE will soon give an updated report on world energy supplies. This report could easily find that there are even smaller supplies of these energy sources, given a decade of increased global consumption and the growing economies of China and India.

If global energy policies and economic priorities do not change, political and social tensions will mount over the supplies and locations of fossil fuels, as they become scarcer and more expensive. The push in Washington for more oil and gas tax breaks is counterproductive and delays developing a carbonless economy. When these measures and others related to a short-term balancing of the budget are implemented, the result will be that future generations will be paying taxes for years to continue 2IR technologies. This short-term fix is misguided, wasteful, and economically crippling for our children and grandchildren. Any new funds and resources must be focused on conservation, reuse, renewable energy generation, and technologies for energy storage, smart green communities, and related areas. A simple first step would be to tax the carbon fuels and use that money to invest in renewable energy power generation and its supporting technologies.

America can no longer afford oil wars, considering their financial impact and their costs in human lives and injuries. Nor can the country afford another environmentally crippling deep ocean oil spill, such as the April

2010 British Petroleum spill in the Gulf of Mexico. And it cannot continue to deny that the nation needs to take a new path.

The United States must establish a national energy policy that makes sense, as the entire country moves rapidly from the 2IR that dominated the 20th century, to the GIR that will be the new world order of the 21st and 22nd centuries. The 2IR was dependent on fossil fuels and internal combustion engines, as well as on heavy manufacturing based on cheap oil and massive infrastructures to support energy and transportation. The GIR is about using renewable energy to power green local communities where renewable power and smart grids can monitor power and increase efficiencies.

Europe, Japan, and South Korea are well on their way to developing sustainable communities and becoming energy independent while a similar large-scale effort is underway in China. In 2008 the Climate Group, an international think tank, reported China's rapid gains in the race to become the leader in developing renewable energy technologies (Jha 2008).

Germany was the number one producer and installer of solar panels for homes, offices, and large open areas from 2006–2009. However, in 2010 Italy held that distinction. By the end of 2011, China will be the number one solar panel and photovoltaic (PV) manufacturer and installer. Japan is now leading the world in auto manufacturing, since it started to make vehicles that are not damaging the environment and atmosphere. Other nations in the European Union, such as the Nordic countries and Spain, have been aggressively implementing policies and programs to become energy independent through renewable energy by 2050. They are succeeding. Denmark has made extraordinary accomplishments already. The Danish government has implemented a national policy, as well as local plans and financing, for locally distributed or onsite energy-renewable power systems.

America definitely has some catching up to do. The sooner it starts, the faster it can achieve the inherent benefits of a sustainable and localized-energy-generated lifestyle, which focuses on sustainable communities, while creating new companies, careers, and areas for employment.

America Must Move toward a Carbonless Economy

The 2008 economic collapse created much of the political confusion and debate in the national dialogue about the environment. The confusion was exacerbated by the American media, which is focused on the businesses of the 2IR. The public is besieged by concepts like sustainability and renewable energy; trends such as "green" energy, jobs, and careers; goals such as energy efficiency and conservation; and threats such as greenhouse gases, global warming, and climate change—all without definitions. California faced the confrontation between the 2IR and the GIR in its November 2010 elections that featured candidates representing both paradigms.

Proposition 23 in California, promoted by various 2IR interests, threatened to reverse the environmental protection laws of the state, claiming that such laws hurt the state's economy. Two Texas oil companies funded the proposition (Environmental Defense Fund 2010). The American Chamber of Commerce broadcasted a continuous mass media campaign supporting the proposition by arguing that the state was losing manufacturing businesses and jobs due to the environmental protection laws. Luckily for California and the nation, the proposition failed. California also got Jerry Brown, a progressive governor who was not beholden to the 2IR. In fact, Governor Brown was a key advocate for the GIR when he was California governor in the late 1970s. As governor again in 2011 for four years, he is taking the state into the Green Industrial Revolution before the rest of the United States.

Part of the problem is defining concepts and ideas such as "clean." For example, there is a qualitative and quantitative difference between "clean" and "green." In 2006, when Al Gore's film *An Inconvenient Truth* was making the public and policymakers aware of the problem of global warming, too many people took the concepts that then were "green-washed" and passed off as something they were not. Thus, natural gas was labeled "clean," and "clean coal" was promoted as better for the environment. Indeed, these fossil fuels are better than oil, but they still have particulates that pollute the environment and cause climates to change. The terms get tossed back and forth by politicians so that everyone thinks they know what they mean, until they try to use them in a sentence. Scientists, too, are drawn into this confusion by accepting grants and positions supported by 2IR fossil fuel companies.

Distinctions must be made between the definitions and meanings of words, concepts, numbers, and even symbols that are often misused (Clark and Fast 2008). The issue is that numbers, words, and ideas are all too often not defined or even discussed. The decision makers just use them, in many cases, to intentionally confuse the public. For example, companies and lobbyists refer to "clean energy," which to them means the use of energy and fuels such as natural gas and diesel. These are fossil fuels and emit gases and particulates that pollute the atmosphere, water, and land. Their chemical wastes cause massive health and environmental problems, which are all a legacy of the 2IR. "Green," on the other hand, in the context of energy and fuel, means renewable energy from natural resources like wind, sun, geothermal, and ocean and tidal waves, as well as the flow of water in rivers. These are the energy resources and systems that drive the Green Industrial Revolution.

Whether America is ready or not, this new Green Industrial Revolution is at its doorstep. Federal stimulus money allotted in 2009 included $200 billion earmarked for energy incentives through conservation, loans, and renewable generation programs. This, coupled with crashing local government budgets, particularly in California and New York, has prompted Americans to look in the direction of energy independence and sustainable activities and communities. However, the focus is primarily on conservation and efficiency as ways to reduce the demand for energy. Reduction of energy use is a primary goal but the facts are that the world is becoming more and more dominated by electricity, and in need of power. There must be a far greater focus of funding for renewable energy so that the GIR can get started in the United States.

Reducing energy use is not enough. For example, consider a typical situation of the $2 million annual energy bill for the small town of Benicia, California (population of 30,000), which represents about 5 percent of the budget (City of Benicia, California 2009). Eliminating that expense would allow the city to beef up safety personnel and community services, or provide a buffer for the leaner days ahead. Unfortunately, most of the Federal stimulus funds for energy focused only on efficiency and conservation. There are no funds for small towns to become energy independent and carbon neutral through renewable energy power generation. Incentives and policies are primarily for tax deductions and credits for renewable energy that are all directed at companies and businesses and assume that they make profits. In the current long-term economic downturn, only the wealthy companies and individuals are able to take advantage of these tax incentives.

While conservation and efficiency are a start, renewable power generation is the core need for the Green Industrial Revolution in terms of innovation, technologies, business, jobs, and career growth. The EU, Japan, South Korea, and now China, have become ever more aggressive in renewable energy generation and technologies through government policies, plans, and funds. China is now the number one wind turbine manufacturer and installer in the world (Greenpeace 2011).

Energy independence and the subsequent elimination or reduction of energy bills are part of the new "American Dream," with potential benefits waiting for the United States. As soon as possible, America needs to give up freebasing fossil fuels and embrace a healthier lifestyle, with intelligent development and greater community connectivity. What is crucial is that Americans must see the vision and take action now. There is no time for a plan that starts in 20–30 years. These actions start in local communities. Almost every community has the renewable resources to make itself energy independent and carbon neutral. America must get started and develop a national energy policy with purpose and funding, and then get out of the way.

In 2004 Woodrow Clark II and Ted Bradshaw wrote a pioneering book on the future of energy policy, *Agile Energy Systems, Global Lessons from the California Energy Crisis.* These agile systems combined central energy plants with onsite or distributed power generation from renewable sources. The book concluded by noting that the "new localized energy market place will redefine how integrated resource management is implemented in a public market where private companies can compete in a socially responsible manner" (Clark and Bradshaw 2004). This is precisely what the GIR is all about. Renewable energy—solar, wind, geothermal, wave, and water—either onsite or delivered to homes and businesses, instead of fossil fuels transmitted or piped from long distances.

The goal should be sustainable communities that integrate renewable energy generation and storage technologies with electric and hydrogen transportation, business development, job creation, and social activities. This is sustainable development, or the interaction between a community's infrastructure requirements, economic needs, and social activities for the protection and preservation of the environment. This interaction stimulates business and provides compelling reasons for pursuing and creating sustainable communities.

Japan, South Korea, and Western Europe have set the pace for sustainable and secure communities in the GIR, through the use of their own renewable energy sources, storage devices, and emerging technologies rather than through importing and using fossil fuels. Sustainable and smart agile communities represent a new paradigm. Nations, states, and cities want to control and centralize power and authority. That has been the historical pattern. However, today with the need to meet and address the global challenge of climate change, regional and local level solutions must be developed, along with policy and action.

In the end, agile sustainable communities must develop and implement strategic plans and funding for energy, waste, water, transportation, and telecommunications. Each sustainable community must redefine the traditional central power plants and unsustainable infrastructure systems into ones that use onsite renewable energy, provide for recycling, control waste, conserve water, and create green buildings. Downsizing, and providing back up and redundant power, among other things, are new and different roles for public agencies and private companies. Today, agile, sustainable, and smart communities are necessary for a less polluted environment and a greener world tomorrow. The solutions to global warming and climate change exist now. America must design and implement them.

Most American cities have the potential to implement some, if not all, of these activities. With a little guidance, our communities, colleges, shopping areas, towns, and cities can have locally distributed renewable energy, clean water, recycled garbage and waste, and efficient community transportation systems that run on renewable energy sources for power. America must create a sustainable lifestyle that is free from the carbon-intensive, fossil fuel–based, inefficient centralized energy generation of the 2IR. Instead of lagging behind, America needs to become the leader in a world on the cusp of historic change. But even better, America can figure out the plans, technologies, and financing to become energy independent, while saving millions of dollars and millions of tons of toxic greenhouse gases.

Renewable Energy: From Central Grid to Onsite and Distributed Power

Renewable energy generation is one of those terms that everyone thinks they understand until forced to use it in conversation. Basically, it is a source of energy that is not carbon-based and will not diminish. For example, the sun shines during the day and the wind blows at night. Each needs some form of storage or feedback technology when the wind is not blowing or the sun is not shining. These forms of energy generation are called intermittent and need technologies to provide for "base load" or round-the-clock power generation. Renewable energy is the foundation for a sustainable community.

The most common renewable energy sources are systems that create power from wind, sun, or water; digestive processes that change waste into biomass; and systems that recycle waste for fuel generation. Other renewable sources include geothermal, "run of the river streams," and now, bacteria and algae.

Wind generation is fairly straightforward. Wind has been used as a power source for hundreds of years. Originally, windmills were used to power small machines for processing or for pushing water. Today, a large propeller is placed in the path of the wind. The force of the wind turns the propeller and a gear coupling interacts with a turbine, which generates electricity. The concept of wind generation may be ancient, but technological advances have transformed it. New-generation wind turbines are stronger, more efficient, quieter, and less expensive.

Wind farms harness the energy of dozens, even hundreds, of wind turbines. Turbines can be installed on land or offshore. They can be placed in small communities or even on building rooftops to capture the natural flow of air.

Solar generation systems capture sunlight, including ultraviolet radiation, via solar cells (silicon). This process of passing sunlight through silicon creates a chemical reaction that generates a small amount of electricity. A photovoltaic or PV reaction is at the core of solar panel systems. A second process uses sunlight to heat liquid (oil or water), which is then converted to electricity. A number of communities are now looking into solar "concentrated" systems, in which the sun is captured in heat tubes and used for heating and cooling of homes, buildings, and central power plants. This is a great renewable energy technology for use in water systems and buildings that have swimming pools.

Biomass (biological material from living or recently living organisms, such as plants) can be used to generate energy using a remarkable chemical process that converts plant sugars (like corn) into gases (ethanol or methane). These are then burned or used to generate electricity. The process is referred to as "digestive" and it's not unlike an animal's digestive system. The appealing feature of this process is that abundant and seemingly unusable plant debris—rye grass, wood chips, weeds, grape sludge, almond hulls, and the like—can be used to generate energy.

Geothermal power is created from heat stored in the earth. This heat originates from the formation of the planet, from radioactive decay of minerals, and from solar energy absorbed at the Earth's surface. It has been used for space heating and bathing since ancient Roman times, but is now better known for generating electricity. In 2007 geothermal plants worldwide had the capacity to generate about 10 gigawatts (gW) (10 billion watts) of power, and in practice generated enough power to meet 0.3 percent of the global electricity demand (Geothermal Energy Association 2010; Geothermal Resources Council, http//:www.geothermal.org). In the past few years, engineers have developed remarkable devices, such as geothermal heat pumps, ground source heat pumps, and geo-exchangers, that gather ground heat to provide heating for buildings in cold climates. Through a similar process, they can use ground sources for cooling buildings in hot climates. More and more communities with concentrations of buildings, like colleges, government centers, and shopping malls are turning to geothermal systems.

Ocean waves have power that can be harnessed to create usable energy. That is the concept behind the revolutionary SeaGen tidal power system, which was pioneered in France and Ireland. The French have been generating power from the tides since 1966, and now Electricite de France has announced a large commercial-scale tidal power system that will generate 10 megawatts (MW) of electricity per year (Ocean Power Technologies, http://www.oceanpowertech.com/).

America, particularly the Pacific coastline, is equally suitable for producing massive amounts of energy with the right technology. Ocean power technologies vary, but the primary types are: *wave power conversion* devices, which bob up and down with passing swells; *tidal power* devices, which use strong tidal variations to produce power; *ocean current* devices, which look like wind turbines and are placed below the water surface to take advantage of the power of ocean currents; and *ocean thermal energy conversion* devices, which extract energy from the differences in temperature between the ocean's shallow and deep waters.

Fuel cells are electrochemical cells that convert a source fuel into an electrical current. They generate electricity inside a cell through reactions between a fuel and an oxidant, triggered in the presence of an electrolyte. The reactants flow into the cell, and the reaction products flow out of it, while the electrolyte remains within it. Fuel cells are energy storage devices that can operate continuously as long as the necessary reactant and oxidant flows are maintained. Fuel cells are different from conventional electrochemical cell batteries in that they consume reactant from an external source, which must be replaced. Many combinations of fuels and oxidants are possible but the best for the environment are ones that derive energy from renewable sources. A hydrogen fuel cell uses hydrogen as its fuel and oxygen as its oxidant. Other fuels include hydrocarbons and alcohols. Other oxidants include chlorine and chlorine dioxide.

Bacterial, or microbial, fuel cells use living, nonhazardous microbial bacteria to generate electricity. British Petroleum has made a $500 million investment in this futuristic process, which is now being developed by researchers at the University of California, Berkeley, and the University of Illinois, Urbana. Researchers envision small household power generators that look like aquariums but are filled with water and microscopic bacteria instead of fish. When the bacteria inside are fed, the power generator—referred to as a "biogenerator"—would produce electricity. Ironically, the funding for this technology comes from British Petroleum, the same company that caused the April 2010 oil spill in the Gulf of Mexico that killed 11 people, damaged the Gulf waters, and polluted the coastland, while destroying fishing and tourist businesses (Robertson and Kaufman 2010).

Storage and Intermittent Technologies

While all these power generation and storage systems produce electricity, none is as cheap today as the currently used fossil fuels—coal, oil, and natural gas. However, fossil fuels were not cheap when their use escalated in the late 1890s and became the foundation for the 2IR.

To maximize renewable power efficiency, renewables must be integrated as linked or bundled supply sources compatible with the natural physical characteristics of the locale. Further, these intermittent power generation resources must include storage devices, because the sun is not always shining and the wind is not always blowing. Today, renewable energy technologies and their integrated systems need the same kind of policy and financing support in the GIR that the fossil fuels got in the 2IR.

Storage devices can be either natural, like salt formations, or artificial, like batteries, flywheels, or fuel cells. Once the electricity is collected, these storage devices let you regulate the distribution so you can optimize energy use. Government support for the 2IR (in terms of tax incentives, funds, and even land) must be repeated for the GIR. Incentives for the 2IR must be reduced and applied to the GIR. This tax shift has been very successful in other industries and businesses and can be designed so there is little or no additional tax burden on consumers.

During the 2IR, there were huge costs to develop central-grid power plants to provide power to U.S. communities. Nonetheless, the 2IR meant that the price of fossil fuels for power plants was reduced over time. It took at least three decades to drive the price of electricity down to affordable levels. There is substantial evidence to document how central power plant companies became economic monopolies around the turn of the 20th century. Smaller power systems were merged into larger ones. This was one of Thomas Edison's goals when he first saw the efficiencies and needs for power generated at one location and then transmitted to consumers.

The merging of these power plant systems during the 2IR led to centralized control of the fuels (primarily fossil fuels such as coal, oil, and gas) and then to a consolidation of the large manufacturing and industrial markets. Market consolidation created monopolies, which state governments then had to regulate. Despite litigation over the next decades to separate and control these utilities, the large fuel suppliers and power generators remained the dominant economic business organizations, with control over demand and prices.

In the 1990s deregulation took hold in America, but it was called "privatization" or "liberalization" in the EU. In both regions, the plan was a failure, as it was primarily based on market economics, which without oversight turned into market manipulation, fraud, and loss of service.

A significant strategy has been to modify deregulation. Central power plants still exist, partly owned by government, but far more standards and codes are needed to oversee supplies, costs, and delivery of energy. German and Danish central power plants, for example, have significant government involvement through either partial government ownership or appointed board members.

The change from either extreme (government-controlled or market-central) power system generation has resulted in more local level energy generation, which is critical in a carbonless economy. While some fossil fuels, like coal, oil and gas, are still cheap today due to their 2IR role in central power plants, they are the major American and global atmospheric polluters. The same is true in China, which now has gained the distinction as the world's number one emission polluter (American Council on Renewable Energy 2011). Per capita, America still ranks first, but in both cases, coal is the major problem. If the human and environmental impacts of coal were calculated into its true costs, then the real cost of coal energy generation for power would soar. To address this problem, the new green economy needs the same sort of tax, and funding support or incentives, that the 2IR received over a century ago.

The result of the 2IR was the creation, operation, and maintenance of large, centralized, fossil-fuel based power plants in the early part of the 20th century and then nuclear power plants in the last half of that century. The plants had to be powerful to withstand the degradations over the vast distribution of a central-powered grid system. At each conversion from alternating current (AC) to direct current (DC), electricity loses some power, but there is so much of it at the beginning that it does not matter several thousand miles away at the end. This results in the loss of efficiency in transmission over power lines as well as the constant need for repairs and upgrades.

Not so in the case of the environmentally friendly renewable systems. For best results, energy systems need local renewable power generation and distribution systems, "smart green" local and onsite grids, so electricity does not have to travel far and suffer losses from inefficiencies. An alternative is to hook into a transmission line. This way, the local grid is added to the existing energy distribution system and the transmission line can act as a battery for the renewable energy that needs storage. Some have equated this to a model of the Internet where there is no single area for control over data, or in this case power; rather, it is spread out and localized.

Energy independence will not happen tomorrow. America spent a trillion dollars on the Iraq and Afghanistan wars. It will cost at least that much to turn America into the leader of the GIR. But America does need to move forward to remain competitive. National and international political leadership are compelling us to quickly surpass what has begun in parts of Japan, South Korea, Europe, and especially China.

Toward the First Steps

Fortunately, some states and communities in America are taking the first steps. Consider California, where the world's largest energy efficiency programs are being implemented. The state is taxing the utility ratepayers and pushing that money back into making buildings and facilities more efficient. California is putting about $3 billion into the 2010–2012 energy efficiency cycle. Renewable energy sources such as solar panels are making possible energy savings targets for the years 2012–2020 of over 4,500MW, the equivalent of nine major power plants (Chan 2011; Lo 2011). However, these programs are still controlled and overseen by the central grid utilities. And even more significantly, they do not lead to less demand for energy. The increase in electronic products including electric vehicles will cause even more energy demands from the 2IR central grid.

New York City, which is struggling to hold onto its leadership in the financial world, is facing severe capacity issues, particularly in Manhattan. Taking a page from California, New York has embarked on a similar state policy–directed energy efficiency effort. New York City has enacted programs and policies to curb pollution, including funding the purchase of hybrid cars. Other states like Pennsylvania, New Jersey, Texas, Illinois, and Missouri are also making steps that will lead them into the GIR. The heavy coal burning states in the Midwest, burning coal from the Appalachian Mountains, are in denial mode. They refuse to give up burning coal, due to the lower costs for energy and the fact that the rancid and toxic residue is blowing east and not spoiling their own environments.

While energy conservation and efficiency are first and important steps, complete energy independence and carbon-neutral policies based on renewable energy are within America's technological grasp. Another generation of renewable technologies is coming and it is much better—lighter, thinner, stronger, and cheaper. Wind and solar power, coupled with highly efficient storage devices, smart green grids, and local onsite distribution systems, are coming together. What is lacking is the large national financing and political leadership to make the commitment and push America passed the threshold into the GIR.

To fully benefit, America must end the lethargy and defeatism of the 9/11 reactionary era and regain the technological inventiveness and entrepreneurship that drove the high-tech boom of the late 20th century. If the United States can do this, then the results will be an expanding green job market, economic revitalization, renewable energy sources, and sustainable communities that are carbon neutral.

Chapter 2

The Problem: Climate Change and Its Global Impact

Earth's climate is changing, with global temperature now rising at a rate unprecedented in the experience of modern human society.

—Arctic Council and the International Arctic Science Committee (IASC) (2004)

That we will run out of oil is no longer in question. The evidence is clear and every major company and nation that produces oil and gas is concerned. The Norwegians recognized and documented the problem in the North Sea in the late 1980s. Today, the transition from 2IR to GIR would not be so necessary if it was simply a matter of delaying when the last drop of the world's oil will fall from the spigot. And it's not just a desire for cheap fuel as much as it is a concern for what the use of cheap fossil and carbon-based fuel is doing to the planet. As scientists have known for more than a decade now, areas of the planet are getting hotter or colder each year where the climate had not changed in thousands of years.

Fly into Los Angeles, Mexico City, Beijing, or nearly any of the world's major cities and you will see the noxious and toxic layers of smog that now surround these communities and cause health problems for their residents. The smog layer has been developing for decades, since fossil and carbon fuels became the major source of cheap energy for the development that has swept the globe. However, as the film *An Inconvenient Truth* documented, this process has accelerated since the turn of the 21st century. The world is round, so what happens in our atmosphere and ocean in one area impacts other regions around the world.

That the residents of these smog-covered communities tolerate this oppressive environmental degradation is a tribute to human adaptability, mixed with a heavy dose of political ignorance. Their adaptability is most likely coupled with the extraordinary desire for cheap personal transportation. It is also a tribute to the monumental skill of the advertising industry, which has convinced the world that a personal vehicle with a 300-horsepower engine that can exceed 80 miles per hour is something to be desired, bought, and celebrated.

Given the overwhelming social and psychological needs of human beings to emulate each other, and America's skill at exporting its lifestyle for profit, it was inevitable, as nations and people struggled to rise from poverty to a middle-class prosperity, that the first thing they did after adding animal protein to their diet was to buy a car and chase the "freedom of the open road." Of course, with almost seven billion people owning one billion cars on an environmentally fragile planet, eventually a price has to be paid. Unfortunately, that price, which started out as the smog blanket common to most major population centers, has now become global warming with potentially disastrous consequences on Earth's climate.

China is now experiencing, in one or two years, what took the Western world a decade or two. In every city in China, personal car ownership has dramatically increased three- or four-fold over the last two years ("Car Ownership in China" 2010). However, the Chinese see and recognize this problem and are aggressively moving to stop and reverse this trend. Their current 12th Five Year Plan (March 2011) is predicted on such national goals (Lo 2011).

The recognition that excessive greenhouse gas emissions are causing global warming and climate change is not new. Though initially advanced by a few visionary scientists as early as the 1980s, and introduced to the world stage with the United Nations Intergovernmental Panel on Climate Change, this realization was sidetracked during America's environmentally insensitive George W. Bush administration. America, under Bush, refused to sign the Kyoto Protocol and derailed international momentum for a cooperative approach to address the threat. But the stage had been set in the Clinton years of the 1990s, with a 2IR Congress and only Vice President Al Gore willing to fight for global constraints on climate change.

The hope is that the more environmentally and scientifically aware Obama administration will reinvigorate America's understanding of the growing scientific evidence of global warming. In 2009, *Global Climate Change Impacts in the United States*, a comprehensive and authoritative report detailing the science and the impacts of climate change on America, was released by the Obama administration under the U.S. Global Change Research Program. The report involved 13 governmental departments and agencies and its release represented a reversal of the Bush administration's environmental policy (Karl, Melillo, and Peterson 2009).

As further evidence of a policy shift, the U.S. Department of Interior National Park Service released reports detailing the impact of climate change on some of the United States's most treasured natural wonders. According to groups like the Rocky Mountain Climate Organization, the research had been ongoing for years, but was not widely disseminated until the Obama administration took office.

One recent publication from the U.S. Department of the Interior quotes the 2004 Arctic Climate Impact Assessment conclusion that the, "Earth's climate is changing, with global temperature now rising at a rate unprecedented in the experience of modern human society" (Arctic Council and the International Arctic Science Committee [IASC] 2004). The publication continues,

> Today, we hear more and more about the effects of climate change. Scientists tell us there is little doubt that human activities are having a major impact on the atmosphere and ecosystems of our planet. Glaciers and snow packs are melting, stream temperatures are going up, coastal erosion is increasing, and changes in weather patterns are leading to drought and heat waves both locally and regionally. According to researchers, the magnitude and pace of these changes, as well as additional ones that climatologists believe to be probable, are unprecedented in human history. (U.S. Department of the Interior, National Park Service 2004)

Nonetheless, there is still significant political resistance to taking action, even when the dangers are clear and immediate. After the BP oil spill in the Gulf of Mexico, the Obama administration pulled back on its initial two-month moratorium on offshore oil drilling. In short, the 2IR appears to be alive and well in America today.

What Is Climate Change?

Since the First Industrial Revolution in the late 1700s, humans have been moving toward large-scale manufacturing. At first, machines started to replace manual labor, horsepower, and wind and water power. Starting in Britain, this transition spread through Europe and eventually reached North America. Societies based on trade and agriculture that were dependent on tools and animals began to rely more and more on machines and engines.

The First Industrial Revolution (1IR) was propelled by the steam engine invented by James Watt. Watt's engine converted the chemical energy in wood or coal to thermal energy and then to mechanical energy. Its main purpose was to power industrial machinery and steam locomotives.

Wood is scarce in Britain (by the end of the Napoleonic wars, most of Britain's magnificent oak trees had been cut into planks and masts for the naval war ships that held Napoleon at bay until his eventual defeat at Waterloo), but the country is blessed with an abundance of coal, which contains twice as much energy as does wood. Coal quickly displaced wood as the fuel for the steam engine. Soon, it was used to produce heat for the industrial processes, to drive engines, and to create propulsion, as well as to warm buildings.

With the help of Ben Franklin and his kite, people began to understand, harness, and commercialize electricity, which had been studied since 1600. The 1IR gave way to the 2IR, and power and energy shifted from steam and coal to oil and the internal combustion engine. (Oil was used in the mid-1800s, several years before electricity. It was burned as kerosene in lamps and small stoves, replacing whale oil.)

As the 2IR took hold, manufacturing grew and the assembly line made it possible to mass produce automobiles. Except for some brutal wars in Europe, Russia, and Asia, and the relentless genocide of the world's aboriginal natives, the overall planet's population expanded. With this expansion came greater distribution of products and increased dependence on fossil fuels.

Society evolved to require greater and greater amounts of energy for light, heat, locomotion, mechanical work, and communications; and then for smart phones, computers, televisions, microwaves, washing machines, coffee makers, and all the other technology and gadgets that make modern living, modern. Since the 1IR, these energy requirements have primarily come from fossil fuels that emit carbon dioxide (CO_2).

Fossil fuels opened the world to the wonders of the personal transportation device. At first, fossil fuels allowed for the transition from an agrarian society to an urban one and provided a way to make electricity. But when used to fuel a car, fossil fuels allowed urban workers to leave their city apartments and settle in the suburbs. This led to the need to construct highways and build housing developments, and created the foundation for America's car culture to take root. Not only did the car become the means to get to work and how we measured success, but also, ironically, it became a symbol of rebellious freedom, like the spontaneous romantic exhilaration of Jack Kerouac's *On the Road*.

Today's new cars are a stunning testament to the creativity and brilliance of human design and engineering. Up close, a glistening, Italian-styled

Ferrari or an impregnable Land Rover are machines that take one's breath away. They are awash in expensive and luxurious leather, encapsulated in a metal shell that is cushioned by multiple protective airbags. The audio systems are transcendent and the navigation, talk-through, and computer systems are the best electrical engineering geniuses can design. All would be better in terms of the environment and the costs to run the vehicles if these extraordinary machines did not run on gasoline.

Unfortunately, since the birth of the 1IR, the Western world's improving lifestyle—and the human passion for autos—has been dependent on fossil fuels. At one time, the coal, oil, and natural gas that powered this new economic model and the prosperity it entails seemed relatively cheap, inexhaustible, and presumably harmless. More and bigger homes and buildings were built, more concrete poured, more fossil fuels extracted and burned. Frankly, there wasn't much to stop this social and economic juggernaut.

As industrialization led to urbanization, and urbanization to suburbanization, America built national highways and thousands of miles of freeways that circled and interlaced its cities. The more concrete that was poured, the faster the suburbs grew and the American lifestyle was forever changed and thoroughly dependent on fossil fuels. The world's undeveloped nations followed and soon India, China, and South America were building highways and suburbs that sprawled along concrete ribbons, creating congestion, generating pollutants, and producing an atmospheric overhang of smog.

The build-up of greenhouse gases has marched in lock-step with the expansion of fossil fuel use since the 1700s (U.S. Department of Commerce 2011). While primarily composed of CO_2, greenhouse gases (GHGs) also include methane (CH_4) and nitrous oxide (N_2O). Greenhouse gases cannot be touched or smelled. Unlike empty beer cans, plastic bags, or the other garbage that pile up along our roads and rivers, GHGs pile up out of sight, in the Earth's atmosphere above our heads. Visualize all the CO_2 that is released from cars, from coal- and gas-burning power generation, and from the burning and clearing of forests and the deforestation of regions like Brazil or Indonesia. The gases float upward into the atmosphere and wrap themselves like a blanket around the Earth. As more and more are added, the blanket gets thicker and warmer.

As a greenhouse gas, CH_4 is 23 times more damaging than CO_2. Like CO_2, CH_4 is released through industrial processes and agriculture as well as through petroleum drilling, coal mining, and emissions from solid landfill sites. Perhaps the worst generator of CH_4 is Bessie, the favorite neighborhood milk cow. Livestock gas is high in CH_4 as well as N_2O. Once CH_4 is released into the atmosphere, it traps heat at a much greater rate than does CO_2. Livestock, and particularly cattle, release both CO_2 and CH_4 through belching as they chew their cud. Climate researchers estimate that the average cow releases about 600 liters of CH_4 per day (Food and Agriculture Organization of the United Nations 2006). N_2O is released through livestock defecation.

According to a United Nations Food and Agriculture Organization (FAO) 2006 report, *Livestock's Long Shadow–Environmental Issues and Options,* the world's livestock sector generates 18 percent more greenhouse gas emissions (as measured in CO_2 equivalent) than does transportation. Livestock are also a major source of land and water degradation. Says Henning Steinfeld, chief of FAO's Livestock Information and Policy Branch and senior author of the report: "Livestock are one of the most significant contributors to today's most serious environmental problems. Urgent action is required to remedy the situation" (Food and Agriculture Organization of the United Nations 2006).

The report notes that the livestock sector accounts for 9 percent of CO_2 deriving from human-related activities, but produces a much larger share of even more harmful GHGs. The livestock sector generates 65 percent of human-related N_2O, which has 296 times the global warming potential (GWP) of CO_2. Most of this comes from manure. And it accounts for 37 percent of all human-induced CH_4 and 64 percent of ammonia, which contributes significantly to acid rain (Food and Agriculture Organization of the United Nations 2006).

About 30 percent of the Earth's surface is now given to livestock production, including 33 percent of the global arable land used to produce feed for livestock, the FAO report notes. As forests are cleared to create new pastures, livestock production is a major driver of deforestation, especially in Latin America, where, for example, some 70 percent of former forests in the Amazon have been turned over to grazing (Food and Agriculture Organization of the United Nations 2006).

Unfortunately, the world's rising middle class wants animal protein (a goal perhaps only second to having a car). Many people believe that a diet rich in animal protein is important for children's growth and mental development. With increased prosperity, people are consuming more meat and dairy products every year. Global meat production is projected to more than double, to 465 million tons in 2050, while milk output is set to climb from 1,043 million tons in 2006. This rapid growth is extracting a huge environmental price. *Livestock's Long Shadow–Environmental Issues and Options* warns, "The environmental costs per unit of livestock production must be cut by one half, just to avoid the level of damage worsening beyond its present level."

Livestock are not the only source of N_2O. The gas is also released as a byproduct of warming temperatures. According to scientists studying the impact on global warming on the Arctic and the surrounding areas of permafrost, warming temperatures are causing N_2O to leak into the atmosphere. A 2009 study from the Arctic Monitoring and Assessment Program, a scientific body set up by the eight Arctic Rim countries, says that the Arctic is responsible for up to 9 percent of global N_2O emissions.

Scientists estimate that there are 1.5 trillion tons of carbon locked inside icebound Earth, mostly in Alaskan and Russian permafrost areas (Arctic Monitoring and Assessment Program 2009). Since the age of mammoths—about 10,000 years ago—N_2O from the carbon has slowly seeped into the atmosphere via lakes and rivers. Over the last few decades, as the Earth has warmed, the icy ground has begun thawing more rapidly, accelerating the release of methane.

Pioneering Work

No one has caused as violent an upheaval in the formerly staid science of climatology as James Hansen, director of the National Aeronautics and Space Administration (NASA) Goddard Institute for Space Studies in New York City. Hansen, an Iowa-born physicist and astronomer, joined NASA in 1967 after doing graduate work at the University of Iowa. At NASA, he gradually focused on planetary research that involved trying to understand anthropogenic (human-made) impacts on the Earth's climate.

Much of Hansen's early work for NASA involved radiative (electromagnetic radiation) energy transfer in planetary atmospheres. This included the interpretation of remote or satellite sensing of the Earth's atmosphere and surface. (Satellites are the most effective means for monitoring and studying global change.) His work included developing global circulation models to help understand the observed climate trends and evaluate human impacts on climate.

In the 1970s, Hansen's published research on Venus proposed that the planet's hot surface was the result of naturally occurring aerosols trapping the planet's internal energy. According to this theory, several billion years ago, Venus's atmosphere was much like Earth's. There were substantial quantities of water on the surface, but a runaway greenhouse effect caused by the evaporation of that water generated critical levels of greenhouse gases in its atmosphere (Hansen and Matsushima 1967).

Hansen's Venus research eventually led him to use the same computer model to understand Earth's atmosphere. He used this model to study the effects that aerosols and trace gases have on the climate. Hansen also developed models that were central to understanding climate change on Earth. Then, in 1988, the Senate Energy and Natural Resource Committee invited Hansen and other scientists to testify on climate and environmental concerns. They, and probably NASA, were not expecting the mild-mannered Midwesterner's blunt assessment.

To the consternation of the Senate committee and to the shock of an unaware national media, Hansen told the senators that, in the first five months of 1988, Earth had been warmer than any comparable period since measurements began 130 years ago. Hanson went further, making the claim that the higher temperatures were attributed to a long-expected global warming trend linked to pollution (Shabecoff 1988).

Until the hearing, scientists had been cautious about attributing rising global temperatures to the so-called greenhouse effect caused by pollutants in the atmosphere. Hansen was not shy and he told the senators that it was 99 percent certain that the warming trend was not a natural variation but was caused by a buildup of CO_2 and other gases in the atmosphere.

Hansen's remarks to the Senate were captured by the *New York Times*. He said that there was no magic number that showed when the greenhouse effect would actually start to cause changes in climate and weather. But he added, "It is time to stop waffling so much and say that the evidence is pretty strong that the greenhouse effect is here" (Shabecoff 1988). Hansen and the other scientists noted that humans, by burning fossil fuels and other activities, have altered the global climate in a manner that will affect life on Earth for centuries to come (Shabecoff 1988).

Hansen's testimony, while not as politically charged as Galileo's defiance of Italian church doctrine in 1615, was nevertheless stunning testimony from someone who worked for such a high-profile government agency as NASA. The testimony was not just controversial; it made him a lightning rod for skeptics, deniers, and protectors of 2IR technologies.

Despite this backlash, he managed to continue his work and further his advocacy to limit the production of greenhouse gases. On several occasions, Hansen's actions have led to his arrest. He was an outspoken critic of public policy under the Clinton and Bush administrations, and was particularly critical of the coal industry. In a 2007 testimony before the Iowa Utilities Board, he said coal contributes the largest percentage of CO_2 into the atmosphere and has called for phasing out coal power completely by the year 2030 (Hansen 2007).

While Hansen introduced climate change to an unaware 1988 America, other scientists across the world were equally concerned with the issue. Reacting to international pressure (that included the United States at the time), the United Nations (UN) established the Intergovernmental Panel on Climate Change (IPCC) in 1988 with the mandate to review and assess the most recent scientific, technical, and socioeconomic information produced by scientists hoping to understand climate change.

The Epic Step Forward: The UN IPCC and FCCC

The world's increased awakening to climate change (and the book you are reading) would not be possible without the United Nations Intergovernmental Panel on Climate Change (UN IPCC). Established in 1988, the UN IPCC was the stepchild of two UN organizations, the World Meteorological Organization (WMO) and the United Nations Environment Programme (UNEP). That the UN IPCC ended up sharing the 2007 Nobel Peace Prize with former U.S. vice president Al Gore is an extraordinary victory of science over politics from the 2IR political and economic interest groups.

The UN IPCC does not do original research or climate change monitoring. Instead, it acts like a scientific clearinghouse, or respected "vetting" panel, that publishes reports on relevant climate change science. A main activity of the UN IPCC is publishing special reports on topics relevant to the implementation of the UN Framework Convention on Climate Change (UN FCCC), an international treaty that acknowledges the possibility of harmful climate change.[1] The UN IPCC reports are based on science from worldwide sources and provide a clear view on the current state of climate change and its potential environmental and socioeconomic consequences. In particular, UN IPCC publishes science that examines the nature and risk of climate change caused by human activity. As of 2011 there have been four major UN IPCC reports published, with more scheduled. There is another one due in 2014. The reports, particularly the first two, were decried and criticized by a variety of business and political groups with interests rooted in the 2IR, or the "dirty" economy (see UN Intergovernmental Panel on Climate Change 1990, 1995, 2001, 2007).

The *First Assessment Report*, published in 1990, served as the basis for the UN FCCC. Its executive summary brought a howl of criticism and calls of corrupt and biased science. The summary said that some scientists were certain that emissions resulting from human activities were substantially increasing the atmospheric concentrations of GHGs, resulting in additional warming of the Earth's surface. Further, the report argued that increased

CO_2 was responsible for over half of the enhanced greenhouse effect. They predicted that under a "business as usual" scenario, global mean temperature would increase by about 0.3°C per decade during the 21st century.

The report brought the potentially dramatic impacts of climate changes to world attention and kicked off the initial controversy. The UN FCCC was asked to do a review and published a supplement to the report in time for the 1992 Earth Summit—the UN's Conference on Environment and Development in Rio de Janeiro. The supplement verified the initial report.

The 1995 *Second Assessment Report* (SAR) was equally insistent that GHGs and excessive CO_2 caused climate change. Sadly, it was equally as controversial, because it tried to determine an economic value on human life. Environmental economics value the health impacts of climate change like any other health risk. Some policymakers and scientists objected, commenting on the difficulty of calculating the costs of climate change on human mortality. For example, the value of a statistical life is assessed to be much higher in rich countries than in poor countries.

The UN IPCC *Third Assessment Report* (AR3), published in 2001, continued to outline a world threatened by the reality of climate change. This report got close, but in the end did not specifically blame human activity for climate change. That would come in the *Fourth Assessment Report* (AR4), issued in 2007. However, before it was published, an amendment to the UN FCCC created international headlines and struck a monumental blow to the political credibility and world leadership of America.

Lord Martin John Rees, the president of the Royal Society of London, summed up AR4 best when he said,

> This report makes it clear, more convincingly than ever before, that human actions are writ large on the changes we are seeing, and will see, to our climate. The UN IPCC strongly emphasizes that substantial climate change is inevitable, and we will have to adapt to this. This should compel all of us—world leaders, businesses and individuals—toward action rather than the paralysis of fear. We need both to reduce our emissions of greenhouse gases and to prepare for the impacts of climate change. Those who would claim otherwise can no longer use science as a basis for their argument. ("UK Scientists' IPCC Reaction" 2007)

Key Findings of the AR4

As the UN IPCC issued reports from one through four, they gained credibility and scientific acceptance. The AR4 has been overwhelmingly accepted by the scientific community (Duray 2007). It is an ambitious achievement, the largest and most detailed summary of climate change ever undertaken, involving thousands of authors from dozens of countries.

The major accomplishment of the AR4, and the one that resulted in its being awarded the 2007 Nobel Peace Prize, was that it provided the science to support the fact that human activity is to blame for global warming and climate change. The strength of the report's science was such that the vast majority of the world's science community has now accepted its conclusions. The argument that Hansen, the blunt speaking Iowan physicist, first uttered to the U.S. Senate committee in 1988 was now accepted science. The "who and what" that was causing global warming and climate change was settled and the culprit was humanity.

The AR4 findings included the following:

- Warming of the climate system is unequivocal.
- Most of the observed increase in globally averaged temperatures since the mid-20th century is very likely due to the observed increase in anthropogenic greenhouse gas concentrations.
- Anthropogenic warming and sea level rise would continue for centuries, even if greenhouse gas concentrations were to be stabilized.
- World temperatures could rise by between 1.1 and 6.4°C (2.0 and 11.5°F) during the 21st century and that:
  - Sea levels will probably rise by 18–59 centimeters (7.08–23.22 inches).
  - There will be more frequent warm spells, heat waves, and heavy rainfall.
  - There will be an increase in droughts, tropical cyclones, and extreme high tides.
- Both past and future anthropogenic carbon dioxide emissions will continue to contribute to warming and sea level rise for more than a millennium.
- Global atmospheric concentrations of carbon dioxide, methane, and nitrous oxide have increased markedly as a result of human activities since 1750 and now far exceed pre-industrial values over the past 650,000 years.

While it is not this book's intention to reproduce this extraordinary publication, here are a few of its many remarkable observations that should convince even the most resistant skeptic that climate change is real, and is a growing environmental and health threat:

Changes in the Atmosphere

Carbon dioxide, methane, and nitrous oxide are all long-lived GHGs. "Carbon dioxide, methane, and nitrous oxide have increased markedly as a result of human activities since 1750 and now far exceed pre-industrial values" (UN Intergovernmental Panel on Climate Change 2007).

Planet Temperature

Cold days, cold nights, and frost events have become less frequent. Hot days, hot nights, and heat waves have become more frequent.

Additionally:

- Eleven of the 12 years in the period (1995–2006) rank among the top 12 warmest years in the instrumental record (since 1850, toward the end of the Little Ice Age).
- Warming in the last 100 years has caused about a 0.74°C increase in global average temperature.
- Observations since 1961 show that the ocean has been absorbing more than 80 percent of the heat added to the climate system, and that ocean temperatures have increased to depths of at least 3,000 meters (9,800 feet).
- During the past century, average Arctic temperatures have increased at almost twice the global average.
- Average Northern Hemisphere temperatures during the second half of the 20th century were very likely higher than during any other 50-year period in the last 500 years and likely the highest in at least the past 1300 years (including both the Medieval Warm Period and the Little Ice Age).

Ice, Snow, Permafrost, Rain, and the Oceans

The report documents increases in wind intensity, decline of permafrost coverage, and increases of both drought and heavy precipitation events. Additionally:

- Mountain glaciers and snow cover have declined in both hemispheres.
- Losses from the land-based ice sheets of Greenland and Antarctica have very likely contributed to sea level rise between 1993 and 2003.
- Ocean warming causes seawater to expand, which contributes to rising sea levels.
- Sea level rose at an average rate of about 1.8 millimeters per year during the years 1961–2003. The rise in sea level during 1993–2003 was at an average rate of 3.1 millimeters per year.

Hurricanes

- There has been an increase in hurricane intensity in the North Atlantic since the 1970s, and that increase correlates with increases in sea surface temperature.
- It is likely that we will see increases in hurricane intensity during the 21st century.

Factors That Warm or Cool the Planet

- Warming and cooling effects on the planet are connected with radiative forcing.
- Carbon dioxide, methane, nitrous oxide, halocarbons, other human warming factors, and the warming effects of changes in solar activity, make individual warming contributions (positive forcing).
- Due to the length of time required to remove carbon dioxide from the atmosphere, human-generated emissions from the past as well as the future will continue to contribute to global warming and sea level rise for more than a millennium (UN Intergovernmental Panel on Climate Change 2007).

Since its establishment in 1988, the UN IPCC has risen to the top of the scientific world. By its expansive inclusion of world-renowned scientists, the UN IPCC has avoided the usual political traps of favoritism and bias that have plagued so many United Nations activities.

Now, after four highly credible reports, a Fifth Assessment Report is coming in 2014. This one will include the work of 831 climate experts (chosen from 3,000 nominations) drawn from fields including meteorology, physics, oceanography, statistics, engineering, ecology, social sciences and economics. In selecting the author teams, the UN IPCC stressed the need for regional and gender balance and recognized the importance of involving new and younger authors.

The Kyoto Protocol

An amendment to the AR4 led to the continued ratification of 1997 Kyoto Protocol, named after the Japanese city where the summit took place. This international treaty intended to bring countries together to reduce global warming and to cope with the effects of temperature increases that were unavoidable after 150 years of industrialization. The provisions of the Kyoto Protocol were to be binding on the ratifying nations.

Countries ratifying the Kyoto Protocol agreed to reduce emissions of six GHGs that contribute to global warming: CO_2, CH_4, N_2O, sulfur hexafluoride, hydrofluorocarbons, and perfluorocarbons. The countries were allowed to use emissions trading to meet their obligations if they maintained or increased their greenhouse gas emissions. Emissions trading would allow nations that can meet their targets to sell credits to those that cannot.

Under the Kyoto Protocol, countries formally agreed to reduce greenhouse gases emissions by 8 percent of the 1990 levels by 2012. Then-U.S. president Bill Clinton hailed the protocol as a historic agreement. However, the U.S. Congress failed to ratify it, effectively making the Kyoto Protocol unenforceable. Clinton's support of the Kyoto Protocol was the last act in a long run by the United States as a world leader on serious environmental issues. However, it was Al Gore who (in Kyoto and later in the national presidential election against George W. Bush) continued to push the issue of climate change as a national agenda item.

For a time, the United States played an active part in global environmental protection. Beginning in the 1970s, the United States had led the world in protecting endangered species, oceans, and fisheries. By the 1980s, the United States was heading up efforts to address the problem of the degradation in the ozone layer. The 1987 Montreal Protocol imposed a stringent ban on the production and use of many substances widely used around the world: deodorants, refrigerants, and propellants for aerosol tins. In 1990, the United States brokered amendments to the protocol that allowed India and China to join. Almost every country in the world ratified the protocol, and the hole in the ozone layer began to close. American political clout and creativity played a major part in this success.

This changed when the presidency changed hands. When George W. Bush replaced Clinton as the American president in 2000 at the turn of the 21st century, he refused to acknowledge the Kyoto Protocol. He argued that U.S. economic interests would be threatened. In condemning U.S. inaction, the British Broadcasting Corporation wrote, "[T]he U.S. contained 4 percent of the world's population but produced about 25 percent of all carbon dioxide emissions. By comparison, Britain emits 3 percent—about the same as India, which has 15 times as many people" ("Q&A: The U.S. and Climate Change" 2002).

This was just the beginning of a litany of criticisms leveled at the George W. Bush administration for environmental failures. For example, a 2004 Report from the Environmental Protection Agency was soundly criticized (Hamilton 2007). The European countries, which had forged ahead on cleaning up their environment and adopting energy renewable policies, were particularly strong in their criticisms. The world's environmental lobby and various leaders in the green movement were equally, or in most cases, harsher in their condemnation of the United States and Bush. With more than 70 major countries signing the Kyoto Protocol, it was clear that the United States had taken a position to which the rest of the world was opposed. This intensified when China and the EU began actively cutting emissions.

Today, the Kyoto Protocol is still unsigned by the United States, although the Obama administration has been trying to work out a compromise. In August 2010, 200 nations met in Bonn, Germany, to consider how to extend the protocol and revisit the turbulence associated with the 2009 climate summit in Copenhagen.

After examining the enormous work done by the UN IPCC, as well as tracing the political and economic struggles of the Kyoto Protocol, it is hard not to conclude that science is right and its detractors are wrong. Yet real progress has not been made and the United States continues to drag its heels. We live in an environment that is under assault, on a planet that has been thoughtlessly ravished by people far too careless with natural resources. The excesses of the 2IR have driven us to the brink and the clock is ticking.

Conclusions from the U.S. Global Climate Change Impacts

The comprehensive report *Global Climate Change Impacts in the United States*, released in 2009 by the Obama administration (Karl, Melillo, and Peterson 2009), detailed the science and the impacts of climate change in the United States. The report reinforced the conclusions of the AR4 and included the following conclusions:

1. Global warming is unequivocal and primarily human-induced.

Global temperature has increased over the past 50 years. This observed increase is due primarily to human-induced emissions of heat-trapping gases.

2. Climate changes are underway in the United States and are projected to grow.

Climate-related changes are already observed in the United States and its coastal waters. These include increases in heavy downpours, rising temperature and sea level, rapidly retreating glaciers, thawing permafrost, lengthening growing seasons, lengthening ice-free seasons in the ocean and on lakes and rivers, earlier snowmelt, and alterations in river flows. These changes are projected to grow.

3. Widespread climate-related impacts are occurring now and are expected to increase.

Climate changes are already affecting water, energy, transportation, agriculture, ecosystems, and health. These impacts are different from region to region and will grow under projected climate change.

4. Climate change will stress water resources.

Water is an issue in every region, but the nature of the potential impacts varies. Drought—related to reduced precipitation, increased evaporation, and increased water loss from plants—is an important issue in many regions, especially in the West. Floods and water quality problems are likely to be amplified by climate change in most regions. Declines in mountain snowpack are important in the West and Alaska, where snowpack provides vital natural water storage.

5. Crop and livestock production will be increasingly challenged.

Agriculture is considered one of the sectors most adaptable to changes in climate. However, increased heat, pests, water stress, diseases, and weather extremes will pose adaptation challenges for crop and livestock production.

6. Coastal areas are at increasing risk from sea-level rise and storm surge.

Sea-level rise and storm surge place many U.S. coastal areas at increasing risk of erosion and flooding, especially along the Atlantic and Gulf Coasts, Pacific Islands, and parts of Alaska. Energy and transportation infrastructure and other property in coastal areas are very likely to be adversely affected.

7. Threats to human health will increase.

Health impacts of climate change are related to heat stress, waterborne diseases, poor air quality, extreme weather events, and diseases transmitted by insects and rodents. Robust public health infrastructure can reduce the potential for negative impacts.

8. Climate change will interact with many social and environmental stresses.

Climate change will combine with pollution, population growth, overuse of resources, urbanization, and other social, economic, and environmental stresses to create larger impacts than from any of these factors alone.

9. Thresholds will be crossed, leading to large changes in climate and ecosystems.

There are a variety of thresholds in the climate system and ecosystems. These thresholds determine, for example, the presence of sea ice and permafrost, and the survival of species, from fish to insect pests, with implications for society. With further climate change, the crossing of additional thresholds is expected.

10. Future climate change and its impacts depend on choices made today.

The amount and rate of future climate change depend primarily on current and future human-caused emissions of heat-trapping gases and airborne particles. Responses involve reducing emissions to limit future warming, and adapting to the changes that are unavoidable.

There Are So Many More Human Beings

Climate change and global warming would not have such an immediate and debilitating impact if fewer people inhabited the planet. Climate change is a matter of scale. A few tons of GHGs won't mean much to the ice cap. The concern comes from the accelerating growth in population and the exponentially increasing number of people entering the middle class.

There are many more human beings today, and people want more of everything. The growing middle classes of the developing world have their eyes squarely on America. They want animal protein and the American lifestyle, with all its extravagant and wasteful excesses and labor-saving devices. People moving up in class want gated communities and McMansions, and fast cars to propel them from the suburbs to the inner city. Knowledge and the way out of poverty are a computer and an Internet connection away. Who can blame them?

There are 6.7 billion people on the planet now, and the UN is predicting that by 2053, there will be 10 billion (UN Department of Economic and Social Affairs 2011). Since the Black Death ended in the 1400s, the world's population has experienced continuous growth. Most of the growth took place during the past two and half centuries. The world population reached 1 billion in 1804, 2 billion in 1927, 3 billion in 1960, 4 billion in 1974, 5 billion in 1987, and 6 billion in 1999. The highest rates of growth were seen after World War II, in the decades of the 1950s–1970s.

According to the UN "World Population Prospects" report:

- The world population is currently growing by 74 million per year.
- Almost all growth will take place in the less-developed regions. Today's 5.3 billion people living in underdeveloped countries are expected to increase to 7.8 billion in 2050. By contrast, the population of the more developed regions will remain mostly unchanged, at 1.2 billion. An exception is

the United States population, which is expected to increase 44 percent from 305 million in 2008 to 439 million in 2050.

- During 2005–2050, nine countries are expected to account for half of the world's projected population increase: India, Pakistan, Nigeria, Democratic Republic of the Congo, Bangladesh, Uganda, United States, Ethiopia, and China, listed according to the size of their contribution to population growth.
- Global life expectancy at birth, which is estimated to have risen from 46 years in 1950–1955 to 65 years in 2000–2005, is expected to keep rising to reach 75 years in 2045–2050. In the more developed regions, the projected increase is from 75 years today to 82 years by mid-century. Among the least developed countries, where life expectancy today is just under 50 years, it is expected to be 66 years in 2045–2050.
- By 2050, India will have 1.6 billion people, China 1.4 billion, United States 439 million, Russia 109 million, Japan 103 million, and United Kingdom 80 million. By continent, Africa will have 1.9 billion, Asia 5.2 billion, Europe 674 million, Latin America and the Caribbean 765 million, and North America 448 million (UN Department of Economic and Social Affairs 2011).

Even before the UN IPCC, scientists were concerned about the impact of increasing population on the environment. The 1994 InterAcademy Panel Statement on Population Growth, which was ratified by 58 member national academies, called the growth in human numbers "unprecedented." The InterAcademy Panel noted that many environmental problems, such as rising levels of atmospheric CO_2, global warming, and pollution, are aggravated by the population expansion. At the time, the world population stood at 5.5 billion (InterAcademy Panel on International Issues 1994).

The Earth's climate is changing dramatically each day, unlike anything seen in recorded time. Scientists have made it clear that human activity is the culprit. The planet's population is growing and the Earth's resources are being extracted faster and faster. The 2IR has locked us into a fossil-fueled and carbon-based economic and social structure and not enough leaders are paying attention. The Green Industrial Revolution cannot come soon enough.

Note

1. Co-author Woodrow Clark was a co-editor and co-author for the IPCC *Third Assessment Report* in 2001 and the lead author/editor for the 1999 UN FCCC's first report on climate change, *Environmentally Sound Technology Transfer from Developed to Developing Nations.*

Chapter 3

Asia Leads the GIR: Japan and South Korea

Asian countries, particularly Japan and South Korea, have led the Green Industrial Revolution (GIR), not just in scientific progress, but also in the commercialization of technology. While it took the political transition that ended the Cold War to prompt Europe to open the door to the GIR, Japan and South Korea—and later Taiwan, Singapore, and India—took a different path.

Japan

Japan is an outstanding example of how green technologies can be successfully created and implemented. The history of Japan starts with it being an island nation that existed for centuries using its own natural resources for energy and development. The natural resources were exploited, but because 70–80 percent of the land is mountainous or forested, the development of land for arable commercial, farm, and residential use was limited. Hence, the concept of "no waste" was known as far back as the Middle Ages, and (as just one example) the lack of large numbers of livestock meant that human waste had to be recycled for fertilizer use.

One feature of living on a small island is being close to one another. This created the need to keep noise under control and to protect the local environment. These concerns are part of the philosophical and cultural roots of the Japanese. And among those concerned were the leaders who set policies and programs.

Lucas Adams, Kentaro Funaki, Claus Habermeier, Russell Vare, and Yu Tanabe assisted on Japan. Special thanks on Korea to Professor Jeongin-Kim, Dean, Graduate School of Industry and Entrepreneurial Management from ChungAng University, and RaeKwon Chung, Director, Environment and Development Division of UN ESCAP(Economic and Social Commission for Asia and the Pacific) and Former Climate Change Deputy Ambassador for the Republic of Korea.

Commodore Matthew Perry's Black ships came from the United States to Japan in 1853 to pressure Japanese rulers into opening up their country to the West. The Japanese called the ships *kurofune* or "black." The older ships were the color black but they also burned coal for power and left clouds of smoke. The 2IR had arrived in Japan.

An interesting historical character from the 2IR period is Nakahama Manjiro, a 14-year-old boy who was shipwrecked in 1841 with four friends who were brothers. He was rescued by an American whaling captain, William Whitfield, who took them to Hawaii. He continued on with Manjiro (nicknamed John Mung), finally dropping him off in Fairhaven, Massachusetts (where a Manjiro Museum in his honor was dedicated in the 1980s). The four brothers remained in Hawaii. In 1850, Manjiro went to California for the Gold Rush and was able to get enough money to go back to Japan in 1851. His English education in math, nautical, and sailing skills were valued by the Japanese leaders (*shogun*) and he became a Samurai (advisor) to the *shogunate* government.

In 1860 the Japanese decided to send a delegation and its first ambassador to the United States. Manjiro became the translator on the Kanrin Maru ship that arrived in San Francisco that year and then went on to Washington, D.C., around the southern horn of South America. Manjiro studied military science in Europe during the Franco–Prussian War of 1870. This led to his translating and teaching the Japanese government and military about navigational systems, which, in turn, led to their successful Japanese–Russian War of 1905 (Kawasumi 1999).

Japan was then a leader of the 2IR. The nation began to conquer Korea and then eastern China in the 1930s, all leading to World War II. Then in the 1950s, after almost 300 years of self-imposed isolation from the rest of the world, Japan became more and more dependent on foreign imports of fossil fuels, especially coal, oil, diesel, and then nuclear power plants in the 1970s.

Japan was devastated first during World War II by bombings, which ended with the nuclear bombs dropped on Hiroshima and Nagasaki, then after the war by the decades-long American occupation, which actually provided leadership and support for the GIR. Although the United States did not move toward the GIR at home, it did encourage the following elements in Japan: innovation, government support and leadership of commercial enterprises, and economic models unlike those being practiced in the United States.

Ironically, World War II proved to be a great benefit to Japan as it leveled the nation politically. It introduced Western economic industrialization, which created an opportunity to rebuild the nation in their way that provided for sustainable growth. At the same time, Japan used knowledge from its past and pre-2IR experiences and included its historical concerns for the environment and limited natural resources. For example, auto industry leaders such as Nissan, Toyota, and Honda developed small cars that were fuel efficient, including today's green megatrend hybrid cars. In the near future, these companies will also be bringing to market cars that use hydrogen fuel cells, one of the core technological areas of the GIR.

While the 2IR was based upon fossil fuels that require mechanical engineering technologies and their support systems, the fuel cell is an example of a product from the electrical and chemical engineering sciences in the GIR. Its business and commercial future depends upon expanding scientific and educational training, and making a dramatic shift in academic and job training for future generations, from mechanical to electrical and chemical engineering. This change reduces impact on the environment and has the potential to reverse climate change.

By the late 1980s, Japan was leading the green revolution. Japan, even more than the EU, was aware of the need to protect the environment. Its carmakers knew that the transportation sector accounted for more than half of the pollution in the environment. Hence, its government supported global searches for technologies that would mitigate and reverse vehicle pollution. By this time, many leading Japanese companies had already begun environmental programs that would enable their products to reduce water use and recycle waste.

Today, due to fuel combustion, Japan is responsible for 4 percent of the global CO_2 emissions, which is the lowest percentage of all major industrialized nations. Nonetheless, Japan intends to reduce its emissions by 60–80 percent by 2050. Emissions in Japan are primarily from coal, but the country is moving rapidly toward renewable energy power generation, as well as new technologies, especially since the earthquake and tsunami in March 2011. Oil accounts for 45.6 percent of the total energy supply, with coal and peat at 21.3 percent, nuclear at 15 percent, and gas at 14.7 percent (Funaki and Adams 2009).

Renewable energy sources such as hydroelectric, solar, and wind account for only 4 percent of Japanese energy power generation, but Japan plays a key role in solar power generation (Chan 2011). For most of the first decade of the 21st century, major Japanese corporations such as Sharp dominated the solar power industry. By November 2008, about 380,000 homes in Japan had solar power generation systems (Funaki and Adams 2009).

Japan's goals are to increase solar power generation by more than ten times today's levels by 2020 and 40 times by 2030 (Funaki and Adams

2009). The government is helping to push this growth. In the first quarter of 2009, it provided 9 billion yen ($99.6 million) in tax breaks for consumers, to promote the use of solar energy systems. Because some solar power generators are large systems, companies are encouraged to concentrate in communities that can support large systems. In Japan, the future for sustainable communities can also be seen in the growth of Zero Emission Homes, which are models built by the Ministry of Economy, Trade, and Industry (METI) to advertise Japan's energy efficient and environmentally friendly technologies.

Because Kyoto was the host site and signatory location for the UN protocols, Japan has committed to reducing its CO_2 emissions by 6 percent from their 1990 base. In 2007 then-Prime Minister Shinzo Abe announced the "Cool Earth 50" initiative to achieve this goal through a variety of advanced technologies in many sectors, from increasing the efficiency of coal-fired power plants to improving safety and reliability for nuclear power plants, as well as through reducing and being more efficient in the use of energy itself (Funaki and Adams 2009). In the fall of 2011, a new government took over with an even stronger commitment to renewable energy systems and reduction of nuclear power due to the March 2011 earthquake and destruction of the Fukishima Nuclear Power Plant.

As Japan and other countries continue to move into the GIR, there must be massive shifts in infrastructures such as energy, transportation, water systems, and telecommunications. Japan is replacing its central grid with local onsite power generation, and has (for the first time) created a Feed-in-Tariff (FiT) that is far more aggressive than the German FiT. Policies such as this are needed to make this dramatic shift happen and be implemented even more quickly.

Since the 1970s, Japan has shown leadership in the GIR. People in Japan do not think of the idea of green technology as anything new. Japanese culture and traditions embrace sustainability, but not as some cool new concept that environmentalists came up with in the last few years. For them, it's a matter of survival. The Japanese pioneered the GIR, not because they were smarter or more responsible, but simply because of the nation's need to sustain a large population in a small space. Japan's green revolution is not about big things but, rather, about sustainable small and practical solutions.

A good business example is the Japanese company TOTO, which uses environmentally sound technologies for its bathroom products (http://www.totousa.com). TOTO's waste-reducing water and energy-efficient toilets and sinks are a mainstay in Japanese homes, office buildings, and airports. TOTO sells a bathtub that is insulated like a Thermos bottle.

Japanese people love taking whirlpool baths, but the water does not stay hot for long. By insulating the tub, TOTO was able to satisfy its customers while reducing demand for hot water. *Kaizen*—the Japanese word for improvement—is a key word for the Japanese GIR.

Another example is INAX, a competitor of TOTO's that invented stain-resistant materials for use in constructing toilets. Even if a toilet manufacturer makes a water-saving toilet, water is still needed to clean it regularly. INAX's use of non-stain material helps decrease the amount of water needed for cleaning. TOTO has since developed similar materials to decrease water use.

The technology that has created a new set of issues is nuclear power. For decades, the Japanese people and their government resisted any kind of nuclear power and were prohibited by their World War II treaty of defeat from developing or using such technology for military purposes. Then, America and its Atomic Agency, which became part of the U.S. DOE in the 1970s, began to push nuclear power. Three major U.S. DOE research labs were created after World War II to focus on atomic and hydrogen energy, and other labs provided support.

During the 1970s, nuclear power plants were promoted to developed nations all over the world. France became the international nuclear power plant leader, with over 73 percent of its energy being generated by such plants. Other nations followed. Yet some countries, such as Italy and the Nordic countries (with the recent exception of Finland), decided against building their own nuclear power plants. Russia built its first nuclear plant after the fall of the Soviet Union in the early 1990s. America has continued to support the building of nuclear power plants in selected nations (Gipe 2011).

This effort suffered a significant setback on Friday, March 11, 2011. An offshore earthquake of 9.0 magnitude struck Japan (their largest ever recorded), and then a devastating tsunami came onshore in northeastern Japan. It destroyed communities and killed thousands. The earthquake caused extensive damage to three nuclear power plants in Fukushima. At the time of this writing, the extent of this damage, its environmental consequences, and its impact on the future of nuclear power are not fully known. While Japan has been a leader in the GIR, its basic mistake (which other countries are making, as well) was to bring nuclear power plants from the 2IR into the GIR. The world must now suffer the consequences of the damage to these crippled nuclear power plants.

The massive Japanese earthquake in March 2011 presented many challenges for Prime Minister Naota Kan. What should prove to be the most significant is the rebuilding of a large area of Japan into a totally sustainable region that leads the rest of the nation further into the green revolution. But even more significant for the Japanese would be to use this horrific series of events to shut down the nuclear power plants. Efforts could then be redirected toward continuing to conserve and be efficient in the use of energy, to develop river and ocean wave power, and to organize and install renewable energy systems in every home, building, and vacant area.

South Korea

South Korea is moving even more aggressively into the GIR than Japan. The central government established a Green Growth Task Force in 2009. Before 2020, its greenhouse gas (GHG) emission level is projected to peak and be reversed to early 1990s levels. South Korea targeted 79 percent of its stimulus money to green technologies. This is the highest percentage of any nation and makes South Korea the seventh largest green technology investment in the world. Of the total amount, $24 billion will be targeted to smart grid infrastructures over the next 20 years (Yang 2010).

Since the 1970s, South Korean economic development strategy has shifted from labor-intensive industry to capital-intensive industry. During this period, heavy chemical industries, especially steelmaking, automobile manufacturing, petrochemical production, and the associated machinery needed, were devoted to fast industrial development. Also, the government made a massive financial investment in these industries. After a gestation period, these industries made up a large part of South Korea's exports. After the early 1980s, the South Korean electronics industry became dominant in world markets, and South Korea is now a leading producer and exporter, exceeding Japan (Lee et al. 2010).

The relative importance of key production factors, such as labor, land, and capital, have changed during Korea's economic growth. Korea placed a high value on labor input during the 1960s and 1970s. During the 1970s and 1980s, Korea put more value on capital input. In the 1980s and 1990s, Korea strongly pushed technological innovation. As a result of the technology push during the 1990s, Korea is now a world frontrunner in information technologies and telecommunications, and their mobile and Internet networks have expanded rapidly.

The rapid increase in Internet users has driven Korea's development and secured its position among the most technically advanced countries in the world. Under a 15-year plan called "Cyber Korea 21," the government is focusing on building a solid information technology infrastructure by expanding broadband Internet access. The government has been able to extend the Internet to 144 regions, covering most of the country (Kim 2010).

The technology push was also responsible for enabling Korea to become the first country to commercialize the use of code division multiple access, or CDMA, chipsets in mobile phones (Kim 2010). Korea eventually developed this technology as a world standard (used in 2G mobile phones). The country was also the first to have an effective third-generation (3G) wireless platform, when it started CDMA2000–1x services in 2000, and followed up with CDMA2000–1x EV-DO services in 2002. The most notable efforts came in the form of satellite-based mobile broadcasting, or satellite DMB. The satellite DMB system embodies one of the most advanced receiver technologies of today, beaming video and audio signals from a satellite to handheld receiver devices in vehicles moving up to 150 kilometers an hour.

Korea developed financial problems in the mid-1990s, as did all of Asia. From 1962 to the mid-1990s, the South Korean government implemented five-year economic development plans based on an economic policy of a quantitative growth paradigm. These economic plans were developed on the premise that labor and capital were the key production factors in any country (Kim 2010). High productivity in labor and capital had made extensive economic growth possible, but this past economic growth strategy had resulted in the conflict between economic growth and quality of life, and had led to increased environmental pollution and degradation, as well as poor income distribution.

By the late 1990s, the Korean economy had undergone a rapid industrial restructuring process because of the economic crisis, which began in 1997. Most experts predict that industrial restructuring will continue in the future, since open market pressure is increasing and the role of multinational companies such as DuPont, Nike, and 3M is growing in the world market (Kim 2010).

Since 2000, the world industrial structure has transitioned from a manufacturing industry to a service industry, and then to a knowledge-based industry. During this time, the economic growth policy and strategy of the Korean government has also changed. In 1970 and 1990, "heavy industrial economic policy" was the key economic policy, but after 1990, "capital-liberalization policies" were dominant in economic policy. Throughout the world, accumulated technology levels in the traditional manufacturing industries have reached the top.

Despite this transition, Korea faces challenges. Today, due to a lack of eco-friendly urbanization, unbalanced regional development, and urban planning from two decades ago, the Seoul Metropolitan Area is suffering from traffic congestion and high pollution levels. Korea's energy consumption and emissions constitute 16.7 percent of the world's total energy consumption and emissions. Emissions in Korea are equal to 4.6 TOE (ton of oil equivalent, a unit for measuring energy), which in 2007 ranked them 11th in the 30 Organization for Economic Cooperation and Development (OECD) countries. Thus, compared to other developed countries (except resource-rich countries like the United States and Canada), South Korea spends too much energy per income level. In terms of energy intensity (TOE/GDP), which measures energy efficiency in a country, South Korea measures 0.315, which in 2007 put it sixth among the 30 OECD countries (Kim 2010). That means that more energy efficiency is needed in the current South Korean industrial structure, although their energy intensity and efficiency measures are improving.

The Korean information industry sector now accounts for 40 percent of the country's GDP and 50 percent of its total exports in 2009, an impressive change from 1997, when the sector accounted for just 7.2 percent of GDP (Kim 2010). South Korea plans to inject more than $1 billion in the information technology industry in 2011, focusing on smart phones, cloud computing, and 4G telecommunications. The country is going to lead the market for mobile dynamic random-access memory, or DRAM, chips and active-matrix organic light-emitting diode, which create more value added in the future (Kim 2010).

The Asian Approach and U.S. Reaction

Asia's move into the GIR contrasts sharply with the United States. Asia uses a national planning perspective that can be seen at the local level. This development becomes sustainable and has a completely different perspective, impact, and rationale for all infrastructures at the regional, state, or national level.

Infrastructure systems need to be compatible, yet integrated with one another. For example, renewable energy power generation must be used for homes, businesses, hospitals, and nonprofit organizations (government, education, and others), but also for the energy in personal vehicles and mass transportation, among other infrastructures.

Economics is critical. And how it is defined even more so. The economic rebuilding that will be needed may give Japan the opportunity to create sustainable communities and recreate businesses and industries, as well as to commercialize new technologies.

Sometimes hardship creates the necessary push for solutions. This could be seen soon after World War II and then the Korean War in the early 1950s, and then in 2011, after the Japanese earthquake. The GIR became dominant in South Korea and then India, Singapore, and Taiwan. The People's Republic of China came into the GIR later, at the end of the 20th century. Germany, Japan, and South Korea now produce cars, IT systems, electronics, and high-tech appliances that dominate global markets.

The United States has not followed Asia's lead. America ignored the emerging commercial technological and economic policy programs in Japan and South Korea. In fact, the United States actually resisted environmentally focused vehicles that got over 40 miles per gallon as well as other environmentally sound programs. Instead of serving as the hardship that creates a push, 9/11 became an excuse. With the United States lobbying for the status quo and then "security," political debate and power struggles drove policy as Americans tried to make sense of a post–Cold War era where special interests replaced reason and stalled movement toward a sound domestic economic policy. Instead, the ideological belief in a "market economy" replaced the reality of how government and industry must collaborate and work together. The EU and Asia were doing just that, while the United States became ideological with its standard excuse: let the "market forces" prevail.

Government involvement and public policies have come to be viewed with suspicion and hostility in the United States. This began under President Reagan in the mid-1980s (the trend was actually started by UK prime minister Margaret Thatcher, and the UK is the only EU nation that is still not part of the GIR). This led to deregulating and privatizing infrastructures ranging from energy to transportation. Contrary to the policies and plans of the GIR, government became the "bad guy." Hence, the United States fell far behind the rest of the developed world through the first decade of the 21st century, with private companies destroying American technological progress and influencing environmental and social policies that impact global climate change.

Meanwhile, Europe and Asia aggressively pursued environmental protection through energy conservation and sustainability policies. They had the financial support of their governments and programs that benefited their new economic strategies, growth, and business leadership. The results are measurable and now place the United States in a "catch-up" economic mode. For example, the aggressive German FiTs had from the early 1990s to the end of 2010 created more than 250,000 jobs and made it the leading

TABLE 3.1. Biggest Cleantech Investment by Country (2010)

1. China	$54.4 B	up 39%
2. Germany	$41.2 B	up 100%
3. United States	$34 B	up 51%
4. Italy	$13.9 B	up 124%
5. Rest of EU-27	$13.4 B	up>1%

nation in solar manufacturing and installation in the world (Habermerier 2009) (see Table 3.1).

The policies of the United States, the UK, and other developed nations were rooted in basic neoclassical economic theory. *The Economist* explained the failure of this economic theory in its mid-July 2009 issue by challenging "modern economic theory." Specifically, *The Economist* pointed out how modern economic theory, such as from the Chicago School of Economics, failed to predict or explain the global economy's "deep recession" in October 2008 ("Collapse of Modern Economic Theory" 2009). In short, as *The Economist* raised the issue, "economics is not a science." That conclusion was neither new nor unknown in most of the EU, Japan, and South Korea over the last 20 years (Clark and Fast 2008).

Global (particularly Western) economic theories were never rooted solely in "market forces" as argued by the Adam Smith and other neoclassical economists from the UK and the United States. Instead, the field of economics has other philosophical historical bases. In Europe, there are different philosophical traditions that lead to a variety of economic theories and practices. In Asia, there are hundreds of years of philosophical roots that form the basis for science. Today in China, the economic system is often referred to as "social capitalism" because of its concern with social issues ranging from the environment to renewable energy to economic prosperity for its total population (Clark and Li 2003; Li and Clark 2009, in press).

The field of qualitative economics, for example, provides deeper meanings and understanding of numbers, words, and concepts that are needed to understand and explain the numbers, words, and definitions of analytical economic work (Clark and Fast 2008). Qualitative economics focuses on the understanding and meanings given to terms, numbers, and concepts pertaining to a society and its cultural history. That approach to economics is more similar to the natural sciences, especially in its use of linguistic theory and generative grammar (Clark and Fast 2008). Other philosophical approaches, such as "lifeworld" in Europe and "symbolic interactionism" from the United States (Blumer [1969]1986), were systematically ignored in the attempt by the neoclassical economists to try to make economics into a science (Clark in press), which they were unable to do.

Government support drives the move toward green energy in Asia. Japan has an explicit industrial program to strengthen the competitiveness of its domestic industries. The Japanese government is funding research and development (R&D) collaborations between government, academia, and industry while offering coordinated deployment incentives to achieve price and performance improvements in a suite of technologies that can improve the productivity of domestic industry. Likewise, South Korea is providing billions in R&D funding and credit guarantees to drive private investment in green energy technologies.

Culture, history, and traditions all contribute to any national economic policy. Japan and Korea based much of their economic growth on those traditions, which directly include government involvement. After World War II and into the Korean War, the United States provided leadership but also learned some economic concepts and management programs from Asia that influenced American business growth. The application of these economic ideals is dramatically different from the neoclassical Adam Smith economic paradigm.

The problem is that the contemporary economists cited in *The Economist* do not understand what science is. Their view of economics is rooted in an idealized paradigm that has flawed theoretical and philosophical roots, yet has dominated the field for the last 40 years ("Collapse of Modern Economic Theory" 2009) and resulted in the recession of 2008.

Thus, there are five key elements for the Green Industrial Revolution: (1) energy efficiency and conservation, (2) renewable power generation, (3) smart local and system grids, (4) energy storage through advanced technologies like fuel cells, and (5) education, training, and certification of professionals and programs.

First, communities and individuals need to conserve and be efficient in their use of energy, and protect natural resources such as land, water, and the atmosphere. Second, renewable energy generated from wind, sun, ocean waves, the Earth's heat, water, and biowaste needs to be the top priority. Third, smart grids are needed on local and regional levels that monitor and control energy in real time. Meters should be used to establish base load use so that energy is conserved (systems put on hold or turned off when not needed) and then renewable energy is generated when demand increases.

Fourth, advanced storage technologies such as fuel cells, hydrogen, batteries, regenerative brakes, and ultra-capacitors are needed to store energy from renewable sources. These devices can store energy from wind and the sun, sources that produce electricity intermittently, unlike carbon-based fuel sources, which produce energy constantly. Fifth, education and training are needed to increase understanding of the importance of energy conservation and to provide the trained workers necessary to make GIR a reality.

The GIR must provide support and systems for smart and green communities so that homes, businesses, government, and large office and shopping areas can all monitor their use of natural resources. For example, communities need devices that capture unused and gray (brackish) water, and equipment that transforms waste into energy so they can send any excess power that is generated to other homes or neighbors. These are some of the best practices from sustainable communities (Clark and Vare 2009). Korea and Japan have been creating and implementing GIR strategies like these since the 1970s.

Chapter 4

Europe Joins the GIR

America has lived off its 20th-century glories for too long. America has generated a string of history-changing events—from the overt military victory of World War II to the covert victory of the Cold War, from the eradication of polio to a landing on the moon, and from the globalization of American culture to the ubiquitous PC and the wonders of digital information technology. America's 2IR-funding money machine, fuelled by President Ronald Reagan's adroit cheerleading of "free market capitalism," lulled the nation into inattention. External threats, like the Cold War and then Islamic fundamentalism, and internal threats like the credit derivative meltdown, went unnoticed until they exploded, draining the nation's resources and straining its psychological resilience.

As America struggled to understand the world's new realities at the end of the Cold War, Europe and Asia were experimenting with a different path. The European countries that had spent countless decades and wasted enormous resources on murderous territorial wars—killing 50 million people and marking the first half of the 20th century as history's bloodiest five decades—were about to sheath their swords. The fight had gone out of these perpetually quarrelling nations, and they started to see the world and themselves through a new, collective global environmental lens.

During the 1960s, Europe's historic miasma started to clear. Europeans started to realize that small bickering nations without a common currency, lacking the ability to transport themselves and commercial goods easily between trading partners, and without the ability to agree on even the simplest mutual political and legal concerns (like copyrights and patents) were going to end up as underdeveloped nations subject to the whims of the

---

Thanks to Professor Henrik Lund, PhD, and Associate Professor Poul Ostergaard, PhD, with their colleagues for Danish data in Energy Developmental Planning at Aalborg University Denmark. And for Italy data, thanks to Teresio Asola Faculty, Chairman of ASM, and Alex Riolfo, PhD, Research Assistant, University of Genova.

world's superpowers. Much to the surprise of everyone, including Europeans, the European Union (EU) came out of this new consciousness, crafted by some of the most brilliant and artful politicians in Europe's history.

A series of world events led to the public acceptance and support that culminated in the establishment of the European Union. The end of the Cold War and the reformation of Eastern and Central Europe fostered European cooperation, replacing nationalistic jingoism.

Two contributing events were the oil crises of 1973 and 1979, which were worldwide wakeup calls. However, it seemed that the Europeans, who were suddenly paying $5 a liter for gasoline, were the only ones really listening. The crises spurred renewed oil exploration, and, fortunately, this led to the discovery of the North Sea's nominal but critical supplies. The North Sea oil deposits—which are rapidly depleting—were enough to prevent social disarray while the EU was coming together, and as the search for alternative energy sources began.

A newfound sense of regional cooperation, coupled with severe energy dislocation and the realization that pollutants were overrunning the continent's fragile environment, triggered the so-called green or environment-first political movement. With roots that go back to the student radicalism of the late 1960s and early 1970s, the green movement was a singular European phenomenon that found support from college students and young activists. Eventually, an alliance between the small Green and Radical parties of Holland and Germany in the1990s led to the creation of the European Green Party (EGP). The EGP, formed at a convention in Rome in 2004, is now considered a viable pan-European political party. In the 2009 European Parliament (EP) elections, the EGP won 46 seats. Environmental issues morphed into legitimate political concerns and were subsequently followed by greenhouse gas (GHG) reduction directives and incentives for renewable energy (Rifkin 2007).

As Europe gained clarity of political and economic purpose, it nurtured the birthing of the GIR. The necessary ingredients for a game-changing megatrend had been developing in Europe during the late 20th century. Regional cooperation, the shock of spiking oil prices, resource depletion, and environmental degradation were the drivers—the midwives of the GIR. Politicians were slowly rounding into action, Germany eventually passed the Renewable Energy Laws (EEG), and soon major corporations were investing and innovating, happy to make a profit on new markets and new demand. Today, major European companies like Holland's Vestas (http://vestas.dk/chia), Germany's BESCO and SOLON, and France's Somfy are international in scope, getting positioned for decades of expansion and profits.

2IR Economic Theory Does Not Work in the GIR

America and Europe diverged. One of the issues separating America from the other developed nations was rooted in basic neoclassical economic theory. "Unlike the Americans who felt that economics was rooted in market forces, the Europeans believed market forces alone were not able to address some key social issues." However, market forces alone were not able to address some key social issues.

Europeans were discovering that each nation needed to have a plan to address the basic systems that interact with citizens and the environment. Each nation's leadership had to address infrastructures like energy, water, waste, telecom, and transportation. Without a national policy, there could be no action, no improvement, and no response to environmental degradation. The key is to have each of these components linked and integrated with the others. That way, infrastructure areas can overlap and costs for construction, operations, and maintenance can be contained and reduced. If infrastructures can be constructed, operated, and maintained on the local level while meeting regional, state, and national goals such as carbon reduction, they take on a different perspective, format, and cost structure.

The need for integrated infrastructures is the primary reason why the GIR has a different perspective and rationale at the local level than at the regional, state, or national level. The systems must be compatible and integrated, but in the case of renewable energy, they should be operated independently to maximize their impact.

Consider Today's European Union

The elements needed to enter the GIR require a total and revolutionary shift from fossil fuels to renewable energy generation, although the GIR has deep roots that go back more than a hundred years. Early auto technology included electric cars and hybrids of gas and electric or other fuels, including biofuels. What happened to prevent the proliferation of these environmentally friendly technologies has been well documented,[1] but the explosive growth and intense dependency on fossil fuels quickly established carbon-based fuels and the internal combustion engine as the core drivers for the 2IR.

By the 1990s, the EU nations became aware that there were serious social, political, and environmental issues connected to this over-reliance on fossil fuels and the internal combustion engine. This understanding helped accelerate the GIR in Europe. Helping the EU's awakening were some public policies that emerged after the EU's Commission held a series of conferences, with corresponding reports and funding in the early 1990s (Habermerier 2009).

Policy leaders began to understand that dependency on Middle Eastern and North African oil and gas was a problem. But the issue became acute in the late 1990s with the growing dependency on Russia and its natural gas supplies. Then, in 2007, the EU really got things rolling with new policies and plans, backed with over 3 billion Euro (about $4.5 billion) in funding for technologies and support (Habermerier 2009).

Germany kick-started its efforts with the FiT, which is a government-legislated financial incentive that encourages the adoption of renewable energy. The German FiT was part of their 2000 Energy Renewable Sources Act, formally called the Act of Granting Priority to Renewable Energy Sources. This remarkable policy was designed to encourage the adoption of renewable energy sources and to help accelerate "grid parity," the point at which renewable energy is available for the same price as existing power from the grid. Under a FiT, those generating eligible renewable energy, whether homeowners or businesses, would be paid a premium price for the renewable electricity they produced. Varying tariff rates were set for different renewable energy technologies, based on the resource development cost. By creating variable cost-based pricing, the Germans were able to encourage the use of new energy technologies, such as wind power, biomass-generated power, hydropower, geothermal power, and solar PV power (Gipe 2010–2011).

Other EU nations followed, and then Canada and some U.S. states and cities slowly began to develop similar policies. The FiT triggered the growth of the solar and wind industries and created many new "green" jobs. Despite having a northern European climate, Germany was a world leader in solar energy generation until 2010, when China took the lead (Chan 2011).

Other European countries have similar programs. Denmark for example, will be generating 100 percent of its energy from renewable power by 2050 (Ostergaard and Lund 2010). While working toward that goal, the country has created new industries and careers. One company, Vestas, is now the world's leading wind power turbine manufacturer, with partner companies and joint ventures all over the world. Vestas continues to introduce improved third-generation turbines that are lighter, stronger, more efficient, and more reliable. Vestas also continues to design new systems, including some that can be installed offshore, away from urban areas.

Frederikshavn, in northeast Denmark, is a city of 25,000 people. The town is converting to 100 percent renewable energy by 2015, with a mix of wind, wave, tidal, run-of-river, solar, and biofuel power. They are tying the renewables together with advanced technologies in storage and grid transmission. By 2007 the city had reached 20 percent of its goal, and by the end of 2009 the city had reached 40 percent by installing 12MW of wind turbine power generation offshore. This will reach 25MW in 2015.

The process began with a city council plan, which included the creation of an Energy Town project. Finance and funding came from the expansion of the district-heating grid, which meets 70 percent of the heat demand. The remaining 30 percent is being supplied by biomass boilers and individual systems, with a mixture of solar thermal and electric heat pumps.

Infrastructure change was critical. Transportation, for example, now includes bikes, cars, and buses using biogas, hydrogen, or electrical sources. Geothermal energy and heat pumps became a growing and useful local energy power source. Where methanol production and biogas plants were used, they were constructed for the short term, with the intent of replacing them with more renewable sources of energy.

European Union Policies

Germany, Finland, France, the UK, Luxembourg, Norway, Portugal, Denmark, Spain, and Sweden are on track to achieve their renewable energy generation goals. The Danes are seeking 100 percent renewable energy power generation by 2050, and they will have 50 percent renewable energy generation by 2015. However, other EU countries are lagging, particularly Belgium, Italy, Netherlands, Austria, Ireland, and Greece, as well as the former nations of the Soviet Union (Rifkin 2004).

The various EU countries have widely different resource availability and energy policy stipulations. France and Finland, for example, are heavy backers of nuclear energy. The UK and the Netherlands have gas deposits, although with reduced output predictions. In Germany, lignite offers a competitive foundation for base-load power generation, although hard coal from German deposits is not internationally competitive. In Austria, hydropower is the dominant energy source for generating power, though expansion is limited.

Other EU directives toward energy efficiency and GHG emission reductions impact electricity generation. Many EU members have taken additional measures to limit GHG emissions at the national level. Since the EU-15 (the 15 countries that were members of the European Union before the enlargement on May 1, 2004) is likely to miss its pledged reduction target without the inclusion of additional tools, the EP and the Council of the European Union (Council) enacted a system for trading GHG emission allowances under the terms of Directive 2003/87/EC, October 13, 2008 (Rifkin 2007). CO_2 emissions trading started in January 2005.

In addition to the trade in emission allowances, the Kyoto Protocol provides for Joint Implementation (JI) and a Clean Development Mechanism (CDM) as further means to meet commitments. JI enables an offset for emissions reductions obtained by projects between two industrialized nations. To receive the offset, one business or entity might invest in a project—for example, increasing the efficiency of a power plant—in another nation that has a duty to limit emissions under the protocol. The resulting emission reductions are credited to the country in which the project was launched and may be transferred to the investor.

The CDM concerns projects in which investors from industrialized countries lower GHG emissions in developing countries. Suitable underlying conditions permitting, JI and CDM can offer a cost-effective alternative to expensive emissions reduction measures taken in the home country.

In 2004 the EP and the Council issued a supplement to the emissions-trading directive, giving companies investing in suitable projects tradable allowances for the reductions in CO_2 emissions achieved. This should increase the diversity of low-cost compliance options, leading to a reduction of the overall compliance costs, while improving the liquidity of the market in GHG emission allowances. The results are questionable.

To trade CO_2 emission allowances, initial allocations had to be made. The directive required members to publish national allocation plans (NAPs) in 2004. Each NAP contained the total emission allowances that a member proposed to issue in the three-year period (2005 to 2007) and described how these allowances were to be distributed. The directive required that members issue at least 95 percent of emission allowances free of charge for the period 2005 to 2007, and at least 90 percent for the period 2008 to 2012.

Germany is a good example of what the EU did to advance the GIR. In the early 1990s, the nation started a FiT program that initially was focused on wind power systems. The program was moderately successful and helped grow the wind turbine industry there. However, as the figures below show, in the early part of the 21st century, the focus shifted to PV and solar-generated energy systems.

As Germany became the number one nation in solar power generation, industrial job creation and related economic development grew exponentially. Despite the fact that the weather in Germany is often cloudy, with long periods of snow and rain, the country has achieved what America and other sunnier areas of the world have not.

However, as a result of the FiT, electricity prices in Germany are the most expensive in Europe and are 70 percent more costly than in the United States (Gipe 2011). However, the German coal industry has been shrinking in recent years to comply with EU mandates and to remain competitive in a global marketplace; German companies and homes with solar power generation can sell their excess power back to the grid and, therefore, offset some of the additional costs from the FiT. In 2011, Prime Minister Merkle advanced more renewable energy through extensions of the Feed-In Tariffs. And said no more nuclear power plans with the current ones being shut down in the next decade.

An aging grid structure and the need to increase capacity complicate the EU's policy decisions and mean that the EU must crank up investment in new energy generation. Estimates indicate that to meet demand in the next 25 years, they will need to generate half-again as much electricity as they are now generating (Ostergaard and Lund 2010). This could result in a profound change in the EU's power generation portfolio. Options under consideration include new plants that produce renewable energy from sources such as wind, solar, and ocean power. Only France has advocated an increase in nuclear power plants, while Sweden and other nations (except Finland) have either reduced or eliminated them.

Distributed Energy Generation for Communities with Renewable Power

Originally, when nations electrified their cities and built large-scale electrical grids, the systems were designed to transmit power from a few large-scale power plants to widely scattered areas. However, these systems are inefficient for handling distributed power (energy from many small renewable-power sources). Although some systems will allow for individual households to either buy power from or sell power back to the grid, the redistribution of power from numerous small-scale sources is not yet well handled.

The grid of the future will be agile, combining a central grid with local renewable energy power generation. The term *agile* helps explain what needs to be an energy strategy for the future, in contrast to the debacle of the California energy deregulation, or as it is called in the EU "privatization" or "liberalization." Agile systems must be smart, flexible, and based on the principles of sustainable development. As the Brundtland Report

(1987) said, "as a minimum, sustainable development must not endanger the natural systems that support life on Earth: the atmosphere, the waters, the soils and the living beings." With that definition in mind, a number of European communities sought to become sustainable over the last 20 years.

There are three basic concerns for sustainable and smart communities that must be addressed. First, there should be political agreement on policies and goals for the entire community. This should include a strategic master plan for infrastructure that covers energy, transportation, water, waste, and telecommunications, while including the traditional academic areas of research, curricula, outreach, and assessments.

Second, issues pertaining to the setting of buildings and overall facility master planning must be addressed from a "green" perspective (to include renewable energy as well as recycling and conservation), and incorporate energy conservation and efficient orientation. These structures should be designed for multiple uses by the community. Dense, compact, walkable areas enable communities to offer a range of transportation choices, which leads to reduced energy consumption and environmental protection.

Third, a sustainable smart community is part of a living network that draws residents from a broader region. This is true for a cluster of office buildings, residential housing developments, government offices, and shopping malls. These buildings and areas require communications and transportation, along with water, waste removal, and energy. Integrated strategies, such as those achieved through industrial symbiosis (sharing services, utilities, and resources among industries), can reduce costs, improve energy efficiency, and reduce waste.

Integrated strategies include distributed onsite power systems for urban sustainable areas. These systems contain features more in common with small rural systems than with the traditional large, urban, single-power plant model of the past.

An example of such a system is one created by ASM, a public utility located in Settimo, near the city of Torino, in the Piedmont region of northern Italy. ASM helped create Laguna Verde, an eco-town that includes a Green Tech Park. The project also has economic value, and is based on the concept that energy production should begin with a "dialogue with the environment." Initially, ASM did a detailed analysis of how much energy was being consumed in buildings, districts, cities, and rural territories. Then it used that knowledge to develop sustainable building and infrastructure designs (Asola and Riolfo 2009).

A similar project, called City Life, is taking place in Milan (in the Lombardy region just east of Piedmont). The regions collaborated and shared information because they needed the help of expert designers and scientists and wanted to include historic areas. For Milan, the goal was to regain a primary role in design experimentation, formal research, and technology innovation. The environmental quality in Milan needs to be improved, which proves to be a remarkable effort (Asola and Riolfo 2009). They focused on an old steel milling area that was located near Settimo so that ASM could provide support and guidance for both regions.

The need for new designs for buildings and infrastructures provided an opportunity for alternative energy products, such as electric vehicles, biomass energy through biogas, and hydrogen fuel cells. Companies were created to develop new environmentally sound technologies like hydrogen fuel cells for homes and motorbikes. Other new companies started to produce products, systems, and technologies at the Green Tech Park in Torino. This effort was lead by ASM in Settimo, where they created "isles" of good practices for energy efficiency and innovative technology integration. This included PV systems, thermal solar systems, cogeneration (microturbines, natural gas endothermic engines, hydrogen fuel cells), mini/micro wind systems, ground and water heat pumps, small hydroelectric systems, geothermal systems for electricity, waste and biomass gasifiers, fuel cell backup systems, and stationary hydrogen plants (Asola and Riolfo 2009).

The result was that the Piedmont region (where Torino is located) and ASM, nearby in Settimo, led the GIR in Italy and set standards for the rest of the EU. In particular, the region and cities involved in these projects set a new standard aimed at enhancing good, ecological, energy-sensible practices. They created a regulatory document that outlined the requirements for certifying buildings as "eco-energetic" or "bio-energetic." They also established incentives for good practices, which have become standards for the GIR.

As the Europeans and the Asians are discovering, the green revolution must start in the home. Energy efficiency and conservation must become part of everyday life. The home is also the place to start with other elements of the GIR, such as renewable energy generation, energy storage devices, and new fuels for transportation. Understanding the GIR on a personal level leads directly to the larger community in which people live and work. People in the EU are beginning to connect sustainability with what they do on a personal level and how they work, go to market, or seek entertainment. When it comes to conservation and efficient use of energy, as well as renewable power generation, storage and telecommunications, the issues are the same for the wider community as within a person's home,

The challenge to the United States is to move rapidly in this direction and overtake the EU and Asia. The United States has the innovation and entrepreneurial spirit and resources to meet and exceed that challenge. It must, or it will become a second-class nation as the GIR takes hold globally.

Note

1. See, for example, documentary films such as *Who Killed the Electric Car* (2008) and *Fuel* (Tickell 2009). See also Hawkins, Lovins, and Lovins (1999); and Brown (2009).

Chapter 5

China Leapfrogs into the GIR

While the Green Industrial Revolution surprised America, it did not catch China napping. Starting in the 1990s (while the United States was thinking that cheap fossil fuel would go on forever and 10-mile-per-gallon SUVs were everyone's natural right), China was thinking ahead with its five-year plans. Since the 1949 revolution, Chinese leaders have consciously set in motion 12 five-year plans and polices. The latest is the 12th five-year plan that began in March 2011 (APCO Worldwide 2010). In the past decade, they saw the need to use their latest plans to "leapfrog" the infrastructure and environmental mistakes made by Western developed nations to gain competitive advantages and mitigate climate change. The result is that China has become a global financial leader with its technological applications of the GIR.

The People's Republic of China sidestepped Western market-driven economics and created world-leading growth with its own style of social and government-led capitalism. Initially, China followed the United States in its 2IR. But as the 21st century began, China followed the approach taken by Europe, Japan, and South Korea. As part of what the Chinese called their "social capitalism" economic model, they merged their interest in making money with their concern for protecting their society and the environment.

Over the past two decades, China converted many of its public infrastructure industries into private companies with foreign investors and shareholders. At the same time, the government retained large percentages of ownership. This model is similar to the one used in Germany and the Nordic countries when they converted their infrastructure industries.

Special thanks to Professor Li Xing, PhD, Associate Professor at Aalborg University, Denmark; Jerry Jin, PhD, technology innovation consultant; David Nieh, environmental economist; and ML Chan, PhD, China-USA Joint Ventures expert.

As the world saw with the spectacularly successful 2008 Beijing Olympics, China had successfully leapt into the GIR.

China used the 2008 Olympics to show that it had arrived as a world leader of the Green Industrial Revolution with massive "green" buildings including on its newly built airport and with a start on the reduction of carbon pollution. While much of the criticism about everything from air pollution to individual human rights was justified, China set the stage and demonstrated its leadership for the 21st century. Nonetheless, there has been a cost in terms of human rights as well as the environment. The Chinese central government is fully aware of both issues and has taken aggressive steps to rectify them.

Since the Olympics, China has continued to invest in clean energy technology. In 2009 China led the United States and the other G-20 nations in annual clean energy investments and finance, according to a new study by the Pew Charitable Trusts (2010). In May 2010, it was reported that China used 34 percent of its stimulus funds ($586 billion) for clean technology energy generation. It will have the capacity for more than 100gW of renewable energy installed and operating by 2020 (Pew Charitable Trusts 2010).

In early 2011, China replaced Japan as the second largest economy in the world (based on gross domestic product) behind the United States (Chan 2011). China has shown how the GIR can spur economic growth through industrial and manufacturing expansion and by building high-speed trains, magnetic levitation train systems, subways, housing, renewable energy systems, and onsite power for heating and cooling that are more and more environmentally sound.

In 2010 the United States lost the "distinction" of being the world's worse air polluter. China is now number one ("World: Now No.2, Could China Become No.1?" 2011). However, if the data were calculated on a per capita basis, the United States would still rank first. Given its focus on sustainable energy sources, China will certainly reverse that trend and be well below the United States and other Western nations in a few years. China's next five-year plan will begin to move the entire country into renewable energy, including electric and hybrid vehicles. The Chinese have the money and resources to make these goals not only achievable, but also sustainable.

The benefits of the GIR go beyond mitigating climate change and halting environmental degradation. They include the social benefits such as more jobs, increased entrepreneurship, and new business ventures, all of which have been, and continue to be, enjoyed in Europe (by Germany and the Nordic countries, in particular). There is even recent evidence that

California is seeing small-scale growth as it tries to move into the GIR with new green jobs in an otherwise moribund U.S. economy.

The shift to renewable energy requires a more educated workforce, upgraded labor skills and businesses that can be certified as environmentally responsible for the short and long term. Along with environmentally sound technologies comes a new green workforce that must learn new GIR technologies that range from nanotechnology to chemical engineering systems. China knows this, and understands that the shift to renewable energy will require extensive workforce retraining. Most of China's senior leaders have degrees in engineering or science, unlike other nations whose elected leaders tend to have backgrounds in economics or law. China's emerging dominance in the wind turbine and solar panel manufacturing sectors are good examples of why having leaders with scientific and engineering knowledge can be beneficial.

China's "social capitalist" economic model has also played an important role (Clark and Li 2003). This model requires that foreign businesses located in China have a 50.1 percent or higher Chinese ownership. The government or newly created government-owned companies then become the majority owners of any new venture. The central government sets a plan, and the foreign company enacts it with projects and financing. Additionally, China requires that the profits made by the new ventures be kept in China as reinvestments.

Vestas, the large Danish wind turbine manufacturer, saw early on that China and Asia were large emerging markets. In the early 1990s, Vestas agreed to China's social capitalist business model and established a joint venture in China. The result was that the wind industry and all the ancillary businesses needed to support it (mechanics, software, plumbing, and electrical work), and those needed to install, repair, and maintain it, grew. China is now the world leader in wind energy production and manufacturing. The solar industry in China is already leading a similar business path.

China intends to learn from the West's mistakes as it moves its energy infrastructures into the GIR (Clark and Isherwood 2010). The country plans to do this through a centralized economic policy that is mainly shaped by the Communist Party of the National Government of China (CPC) through plenary sessions of the People's Republic of China Central Committee. The committee plays the leading role in establishing the foundations and principles of Chinese policy by mapping strategies for economic development, setting growth targets, and launching reforms.

Long-term planning (in five-year increments) is a key characteristic of centralized social economies, as one overall plan normally contains detailed economic development guidelines for the various regions. There have been 12 such five-year plans; the name of the 11th plan was changed to "guideline" to reflect China's transition from a Soviet-style communist economy to a social capitalism economy.

In October 2010 China announced the 12th five-year plan with final approval and implementation plans passed in March 2011. The 12th plan ends in 2015. The guideline addresses rising inequality and sustainable development. It establishes priorities for more equitable wealth distribution, increased domestic consumption, and improved social infrastructure and social safety nets. The plan represents China's efforts to rebalance its economy, shifting emphasis from investment to consumption and from urban and coastal growth to rural and inland development. The plan also continues to advocate objectives set out in the 11th five-year plan to enhance environmental protection, which called for a 10 percent reduction of the total discharge of major pollutants in five years.

The 12th five-year plan focuses the nation on reducing its carbon footprint, and will address climate change and global warming. Not only will it be well financed, with the equivalent of $1 trillion U.S. dollars, but it will set in motion the possibility that China will be able to surpass Western nations in the technologies and industries that support and make up the GIR. One key element is that the Chinese will change their central power plants into agile, sustainable, and distributed-energy infrastructures with local onsite power systems that use renewable energy to power the facilities.

An Emerging World Leader in Sustainability

There has been a dramatic transformation in Beijing and several other major cities in the world's most populous nation. In the 1990s Beijing was ranked third among world cities with the highest levels of air pollution (Clark and Isherwood 2010). Today, air pollution has been reduced to a level that prompted a Beijing citizen to say, in a voice of full of wonder, "Now we can see the stars at night!" This dramatic change was inspired by Beijing's need to modernize and curb its emissions well before the 2008 Summer Olympics.

Sustainable development is now official government policy in China, and it has been implemented at a remarkable pace in some regions. Shanghai has built the first commercial high-speed magnetic levitation, or maglev, train line, which was based on cutting-edge German technology (the technology was originally licensed from the United States in the 1990s). The Shanghai Maglev train connects their new Pudong International Airport to the city's rapidly growing subway system, making the 30-km trip in just seven minutes. Furthermore, Shanghai, Beijing, Shenzen, Qingdao, and Chengdu are all building or planning new underground railway lines, and in some cases, light rail lines at ground level. High-speed rail now connects many Chinese cities, providing shorter travel times than can be achieved by airplane.

The transformation of Beijing, Qingdao, and Nanjing into postmodern GIR cities was driven in large part by China's goal of making their 2008 Olympics green. For example, PV panels and concentrated solar panels were constructed on many of the Olympic buildings. The determination to achieve sustainable development in other cities like Shanghai, which was not an Olympic city, cannot be explained in this way, but Shanghai did host the World Expo in 2010, which provided an opportunity to showcase its implementation of significant GIR technologies and its plans for the future.

In Shanghai's urban planning exhibition, which attracts thousands of visitors and business people each day, Shanghai states that it wants to become one of the world's leading commercial cities in the 21st century. To do this, Shanghai cleaned up its air and water. As new buildings and sections were added, they installed subways and roads that allowed for bikes and walkways.

The Chinese recognize that sustainable development of an area or region is good for business and tourism, as well as for its citizens. Numerous cities and regions are now labeled as sustainable and have established benchmarks and criteria for official certification. Hundreds of conventions and conferences are held in these cities, primarily to show visitors the positive environmental impacts, but also to create business opportunities.

To encourage sustainable development, the Chinese national government continues to strengthen environmental legislation and make huge investments in green technology and sustainable infrastructure improvements. The nation's environmental protection sector is projected to grow at a 15 percent annual rate (Lo 2011). Cities are closing their most polluting factories and moving others to locations far from residential and commercial areas. China encourages industries to modernize, which has improved energy efficiency over the past decade. However, the growing demand for personal cars and new buildings and homes has increased national energy consumption. To meet this demand, the government is building traditional fossil fuel facilities and nuclear power plants, as well as large wind and solar farms, which it will combine with renewable energy onsite and distributed power systems.

China's Energy Needs Are Huge

As the world's most populous country, and one with a rapidly growing economy, China has huge energy needs that will continue to grow in the future. China's GDP grew at an average of 10 percent between 2000 and 2008 (Lo 2011). In 2011 China was ranked Number 2 globally in GDP (Lo 2011). China is also the second-largest oil consumer behind the United States. A net oil exporter in the early 1990s, China is now the world's third-largest net importer of oil. In 2009 China's oil consumption growth accounted for about a third of the world's oil consumption growth (Chan 2011).

Natural gas usage in China has also increased rapidly in recent years, and China has looked to raise natural gas imports via pipelines from Russia and liquefied natural gas, primarily from Australia. China continues to buy and invest in energy-producing companies from around the world.

China is also the world's largest producer and consumer of coal, an important factor in world energy markets, but one that also creates significant water and emissions impacts from drilling, shipping, and burning coal. With 70 percent of its energy derived from coal, another 20 percent from oil, and less than 1 percent from renewable sources, China has a way to go to reach energy sustainability (Lo 2011). Nevertheless, China is intent on reducing its fossil fuel dependence.

Unlike the United States, China has removed most of the subsidies for the production and use of fossil fuels. By 2035 China plans to reduce coal use to 62 percent through increased efficiencies, and reduce its carbon emissions by at least 40 percent from 2005 levels by 2020 (Lo 2011). With its focus on reaping the benefits of the GIR, China plans to increase non–fossil fuel energy consumption to 15 percent of the energy mix in the same time period.

China's dramatic growth and increased energy demands put added pressure on global supplies of fossil fuels and has motivated China to purchase large fossil fuel companies and contracts around the world. In addition, energy demand will have an enormous impact on China's already burgeoning solar and wind energy generation industries. Spurred by the government's social capitalist economic policies, which have resulted in rapid business and job growth in the green industry, China is now the second largest manufacturer of solar panels in the world, and is expected to be the largest in 2011 (Chan 2011). China is poised to lead in advanced batteries, high-speed rail, hybrid and electric vehicles, nuclear, and advanced coal technology.

China's Solar Valley City

China has a strong commitment to solar energy generation. The Chinese have built Solar Valley City in Dezhou, Shandong Province. This ambitious project will create a new sustainable, environmentally sound center for manufacturing, research and development, education, and tourism focusing on solar energy technologies. Solar Valley City is part of China's efforts to promote green energy technology and grow global market share.

More than 100 solar enterprises, including major solar thermal firms, are based in Solar Valley City. The solar industry in China employs approximately 800,000 people, and China's solar thermal industry and the accompanying industrial chain are examples for the rest of the world. A leading company, Himin, produces more than twice the annual sales of all solar thermal systems in the United States (Kwan 2009), and is quickly expanding into solar photovoltaic and other technologies.

The Chinese solar industry (unlike its wind turbine industry) is an export industry. Toward the end of 2010, China became the world's largest producer of PV cells, but because approximately 98 percent of sales of PV solar products were exports, the industry was hit particularly hard by the worldwide financial crisis (Chan 2011). Solar industry leaders have lobbied for a more active set of government policies (similar to those in the wind industry) to subsidize the domestic use of solar power. Because there is so little domestic use of solar power, the potential for growth is strong. Policies intended to jumpstart domestic solar power demand and turn around China's overly export-oriented PV industry are emerging and the Chinese Ministry of Finance is pushing an onsite or local Solar-Powered Rooftops Plan.

The Solar-Powered Rooftops Plan will develop demonstration projects for building integrated solar power (including solar power rooftop units and PV curtain walls) in large and mid-sized cities that are economically developed and want to be sustainable. The plan also supports the development of PV systems in villages and remote areas that are outside the reach of the power grid. As part of this effort to improve domestic use of solar panels, the Ministry of Finance has earmarked a special fund to provide subsidies for PV systems that are at least 50 kilowatts (kW) in size and have 16 percent efficiency. The subsidy will cover the cost of the equipment, or approximately 50 to 60 percent of the total cost of an installed system (Sundra Solar Corporation, Beijing, http://www.sundrasolar.com).

Industry analysts say that much needs to be done to develop a thriving solar industry in China. The country will need to reorient the solar industries from one that relies on foreign trade to one that is balanced between domestic consumption and export. To achieve this balance, the Chinese will need to create a new domestic system that matches the industry's export capabilities.

Wind Power

China's most economically competitive new energy source is wind power. The nation's wind industry emerged in 2005, after a decade of joint ventures and collaborations with northern European companies. Favorable government policies were key to doubling the country's wind power capacity each year. According to the Chinese Renewable Energy Industries Association, China has the world's largest installed wind turbine capacity. In 2010 China had a total wind power capacity of 41.8gW, an increase of 16gW, or 62 percent, from a year earlier.

> Wind was the favorite sector of renewable energy with 94.7 billion USD invested globally (44.1 in developed countries and 50.6 in developing countries), where 38.1 billion USD came directly from China's government investment . China's 12th Five Year Plan places even more focus and funds for wind and solar in the western region of the country to provide renewable energy to encourage sustainable development. (Li and Clark, 2012)

The wind industry is essential to achieving China's goals of secure and diversified energy production. The industry also contributes to economic growth, environmental and pollution control, as well as GHG reductions. The Chinese Renewable Energy Industries Association (CREIA) estimated that in 2009, the wind turbine industry provided an output value of 150 billion Renminbi (RMB), which with taxes and fees paid to support the national finances was valued at more than 30 billion RMB (Chinese Renewable Energy Industries Association, http://www.reep.org).

China also plans to reduce emissions through the use of wind-generated power. If the Chinese wind power industry installs 200gW by 2020 (with a power generation output of 440,000 gigawatt hours [gWh]), it will reduce GHG emissions by 440 million tons. China will also limit air pollution by reducing coal consumption, and CREIA predicts that at the same time the country will generate more than 400 billion RMB in added value and create 500,000 jobs.

While work is needed to integrate renewable energy into the China's electricity grid, the country is intent on a massive increase in wind generation. China is rich in wind energy resources, with a long coastline and a large landmass. Wind energy resources are particularly abundant in the southeast coastal regions, the islands off the coast and in the northern part of the country. The western inland regions are also rich in wind energy potential.

Offshore wind energy resources are also plentiful, and in 2010 the first large offshore project was completed at Shanghai's Donghai Bridge. Thirty-four large 3MW turbines, producing 100MW, were installed. Analysts estimate that as much as 32,800MW could be installed by 2020 (Chinese Renewable Energy Industries Association, http://www.reep.org).

Wind energy has enormous potential in China and could easily become a major part of the country's energy supply. Some scientists estimate that the total capacity for land-based and offshore wind energy could be as high as 2,500gW.

China's wind turbine equipment manufacturing industry has developed rapidly by reaping the benefits of the green revolution. The GIR has provided substantial new business and job growth through the development of green technologies. Domestic wind turbine manufacturers now account for about 70 percent of China's supply market and are beginning to export their products. The largest manufacturers are Sinovel, Goldwind and Dongfang Electric. China now leads the world and accounts for roughly a third of the global total, both in installed wind turbine capacity and in equipment manufacturing capability (Zhao, Hu, and Jian Zuo 2009).

The state-owned power supply companies have developed the largest wind farms. These companies are pushed by national law to steadily increase their proportion of renewable energy. The CREIA reported that by the end of 2009, a total of 24 provinces and autonomous regions in China had their own wind farms, and more than nine provinces had a cumulative installed capacity of more than 1,000MW, including four provinces exceeding 2,000MW. The Inner Mongolia Autonomous Region was the lead region, with newly installed capacity of 5,545MW and a cumulative installed capacity of 9,196MW (Clark and Isherwood 2010).

At the 2009 UN Copenhagen Conference on climate change, China committed that by 2020 it would meet 15 percent of the nation's energy demand with non–fossil fuels. Achieving this goal will require a huge increase in green energy development, including a much greater concentration on wind power.

Through the Renewable Energy Law and other policies, China has made a major commitment to wind energy. A major part of future efforts involves the creation of seven major scale wind power bases. Each wind base has potential for at least 10gW of installed capacity.

The National PRC Energy Bureau is developing these bases. They plan to create a total installed capacity of 138gW by 2020, but only if the supporting grid network is established. A significant problem is that many of these bases are located in remote areas with a weak transmission grid and a long distance from China's main electricity load centers. There are also concerns about how to integrate large quantities of variable wind power into a grid built for coal-burning power stations.

Pricing is another important element. China's support mechanism for wind power has evolved from a price based on return on capital to a FiT, with variations based on differences in wind energy resources.

The FiT system was introduced to China in 2009. The system divides the country into four categories of wind energy areas. This regional FiT policy seems to be a positive step in the development of wind power and is stimulating stronger economic growth, increasing manufacturing output, and adding jobs. Additionally, the Chinese see the need for trained workers for building, operating, and maintaining these new systems, so they have created engineering and science programs to train people to work in wind and other renewable technology industries.

China faces several challenges when it comes to integrating large-scale wind-generated energy into its local and regional grid networks and infrastructures. Wind farms in China are located mainly in areas far from load centers, and where the grid network is relatively weak. This causes a loss in efficiency, so the present design of the infrastructure grids places constraints on the development and use of wind power. This has become the biggest problem for the future development of wind power throughout the country. However, the Global Wind Energy Council projects exceptional growth for China's wind power capacity. They predict it could reach 129gW by 2015, 253gW by 2020, and 509gW by 2030. Wind power would account for 10 percent of total national electricity supply by 2020 and reach 16.7 percent in 2030 (Chinese Renewable Energy Industries Association, http://www.reep.org). These figures do not take into account more local and regional wind farm systems or smaller systems that are integrated into buildings.

The Green Technology Tigers

In a pattern reminiscent of the post–World War II rise of Asia's manufacturing sectors, China, Japan, and South Korea are leading the way in developing green energy technologies. As a result, these countries have been dubbed the "green technology tigers." The three nations have already passed the United States in the production of renewable energy technologies. According to the Breakthrough Institute and the Information Technology and Innovation Foundation 2009 report *Rising Tigers, Sleeping Giants,* these nations will out-invest the United States three-to-one in renewable energy technologies (Bradsher 2010), as confirmed by the latest studies in 2011 (Pew Charitable Trusts 2011). This will attract a significant share of private sector investments in green energy technology, perhaps trillions of dollars over the next decade. Asia's green technology tigers will, therefore, receive the benefits of new jobs and increased tax revenues at U.S. expense.

Government policies and investments are the keys to helping China, Japan, and South Korea gain a competitive advantage over the United States, and even Europe, in the green energy sectors. These Asian nations are making a large direct public investment in GIR technologies. Government investments in research and development, green energy manufacturing capacity, the deployment of green energy technologies, and the establishment of enabling infrastructure will allow them to capture economies of scale, learning-by-doing, and innovation advantages.

While the Asian governments are making large-scale investments, the United States relies on modest market tax incentives that are indirect, create more risk for private market investors, and do less to overcome the many barriers to green energy adoption. Tax breaks make sense when the economy is strong and they work well for large companies. However, during a long global economic recession they pose a barrier to building and operating green energy systems in the United States (especially for smaller companies).

Companies that can establish economies of scale and create learning-by-doing opportunities ahead of competitors can achieve lower production costs and manufacture higher quality products. This will make it harder for new entrants to break into the market. Direct government investments will help Asia's green technology tigers form industry clusters, similar to California's Silicon Valley, where inventors, investors, manufacturers, suppliers, universities, and others can establish a dense network of relationships. Even in an era of increasingly globalized commerce, the structure of these regional economies can provide enduring competitive advantages.

In China, national, regional, and local governments are offering green energy companies generous subsidies—including free land, funding, low-cost financing, tax incentives, and money for research and development—to establish operations in their localities. It took just three years for the Chinese city of Baoding to transform from an automobile and textile town into the fastest-growing hub for wind and solar energy equipment makers in China. The city is home to "Electricity Valley," an industrial cluster modeled after Silicon Valley, that is composed of nearly 200 renewable energy companies that focus on wind power, solar PVs, solar thermal, biomass, and energy efficiency technologies. Baoding is the center of green energy development in China, and operates as a platform that links China's green energy manufacturing industry with policy support, research institutions, and social systems.

In Jiangsu, a province on the eastern coast of China, local government officials have provided large subsidies for solar energy with a goal of reaching 260MW of installed capacity by 2011 (Li 2010). Jiangsu already houses many of China's major solar PV manufacturers, and the new policy is targeted to create substantial market demand and attract a cluster of polysilicon suppliers and solar technology manufacturers.

Another Chinese city, Tianjin, is now home to the Danish company Vestas, the largest wind energy equipment production company in the world. This base not only enhances the company's production capacity, but also increases the number of locally installed wind turbines and helps component suppliers develop expertise with the company's advanced wind power technology, while providing a learning laboratory for students and researchers.

Regionally based programs provide cost and innovation advantages, including access to specialized labor, materials, and equipment at lower operating costs, as well as lower search costs, economies of scale, and price competition. A regional focus provides member organizations with preferred access to market, technical, and competitive information, while creating knowledge spillovers that can accelerate the pace of innovation. Relationships between companies are leveraged and integrated so they can help each other learn about evolving technologies as well as new market opportunities. Workforce mobility enhances the rate of innovation for the whole region, making it both sustainable and part of the GIR. These regional areas provide an attractive business environment for particular industries; if one or two companies fail or move out of the area, others can quickly replace them.

The United States created some of the first examples of these kinds of economic regions. Detroit became a center for automobile production, and its early leadership in auto technology made it a world industry leader for most of the 20th century. Later, Silicon Valley became a center for information technology. Developing regions to promote science, innovation, venture capital, and relationships among organizations provided strong competitive advantages that made it costly for other nations to catch up. But the difference with the GIR is that the U.S. government rarely provides outright financial support other than tax breaks and limited research funds. Conversely, China and other GIR nations have provided proactive and long-term governmental commitments through planning, financial investment, and strong international networks for marketing, sales, and support.

Establishing industrial regions does not guarantee continued market dominance. In the case of the automotive industry, U.S. firms eventually lost market dominance after East Asian nations spent years implementing an industrial policy that sheltered their nascent auto industry from competition. At the same time, these nations invested billions in direct subsidies to support the industry's growth and technological progress. Above all, the Asian nations held a high regard and value for the environment and moved aggressively into hybrid, electrical, and hydrogen technologies. In many cases, they used technologies and innovations licensed from U.S. research and university laboratories. In the face of this dedicated international competition, and its own failure to innovate and adapt, the U.S. auto industry faltered, failed, and two out of three of the major companies—GM and Chrysler—went bankrupt.

To regain economic leadership in the global green energy industry and become part of the GIR, the U.S. energy policy must create a long-term plan that includes large, direct, and coordinated investments in green technology research and development, manufacturing, deployment, and infrastructure. If the United States hopes to compete in the GIR with new green energy industries, it must close the widening gap between what the U.S. government invests and what other nations are investing. The Unites States also must provide more robust support for green technology research and innovation, manufacturing, and domestic market demand.

The United States has many of the elements in place to compete with Asia in the green technology sector of the GIR. As was shown in the 2IR, America has natural innovation advantages, including a skilled workforce, world-class universities and research institutes, robust capital markets, an open society, and a vibrant creative culture. However, given the GIR now advancing in Asian and EU countries in the green energy sector, these historical advantages will not be sufficient for the United States to retain an innovative edge. Recently, the U.S. Information Technology and Innovation Foundation ranked the United States sixth out of 40 leading industrialized nations in innovation competitiveness, and last in the rate of improvement in national innovation competitiveness, over the last decade (U.S. Information Technology and Innovation Foundation 2009). America's economic competitors are surging ahead while U.S. innovation capacity stagnates.

Chapter 6

Sustainable Communities

The icon for modern civilization is the city. Cities such as London, New York, Paris, Rome, and now Moscow are extraordinary, glittering national monuments to civilization that have attracted residents for centuries. Cities are where people in developing nations go to find work, why they give up subsistence farming, and where they can get a paycheck, social services, and, most importantly, schools for their children so they can escape the poverty cycle. To do all this, cities require enormous amounts of energy. It is the GIR that will change cities into sustainable communities.

Today, half the world's population of 6.7 billion people live in cities, mostly as a result of the 2IR. If these trends continue, by mid-century, 75 percent, or about 7 billion people, will live in a city. China already has several cities with more than 10 million residents and regions that have 30–40 million people living, working, and retiring. In 1975 there were five megacities (cities with populations of more than 10 million) around the world. By 1995 there were 14 and, according to the United Nations, by 2015 there will be 26, most of them in China (UN Department of Economic and Social Affairs 2011).

As a planet, Earth has been trending toward ever-larger social organizations since humans gave up the hunter-gatherer lifestyle. The more humans evolve, the less willing they are to tolerate a rural or agrarian existence. The lures of bright lights, easy money, and the rest of modernity are too much to ignore and so the steady march to the cities continues.

While planners, architects, and politicians have watched and profited from this phenomenon, they have allowed most cities, particularly those in America, to grow without regard to available resources or ecological impacts. Why else would a city like Las Vegas be allowed to sprawl unchecked across a fragile desert without apparent concern? There are about two million people living in this water-constricted environment. It's just a question of time until the water runs out.

Los Angeles: A Car-Centric 2IR City

Other large sprawling North American cities, which are best represented by Los Angeles, are equally developed without regard for ecology. These towns line the freeways, one neighborhood running into another, without regard for geography or logic. Greater Los Angeles has a population of approximately 15 million people. It is a region locked in the fossil-fuel 2IR era, a car-centric culture that is dependent on a network of interlaced freeways and highways.

Not only does Los Angeles have the largest high-speed roadway network (known as a freeway) in the world, but it also has the highest per-capita car population in the world. While the United States may be the country with the most registered vehicles (around 232 million), California, and in particular Los Angeles, are the places most committed to and dependent on the automobile—the backbone of the 2IR. The Greater Los Angeles region holds the highest concentration of cars in the world, with more than 26 million, or about 1.8 cars per person. Los Angeles is the world's most car-populated area of urban sprawl in the world.

The Los Angeles freeway system handles more than 12 million cars daily. This helps explain why Los Angeles holds the number-one spot on the list of America's most congested and polluted roadways, according to the 2005 Urban Mobility Report by the Texas Transportation Institute. In recent years, the situation has gotten worse. The availability of new automobiles with low emissions and more miles per gallon has led to increased driving, causing more pollution.

More automobiles and more driving mean more air pollution. The Los Angeles Basin, which is susceptible to atmospheric inversion (a meteorological anomaly that traps pollution and keeps it close to the ground), contains exhaust from road vehicles, airplanes, locomotives, shipping, and manufacturing. Millions of tons of toxins are released daily into the atmosphere. Due to its semi-desert climate, Los Angeles does not get much rain, so pollution accumulates as a dense cover of smog, threatening the health and well-being of the residents. This heavy layer of carbon-intensive pollution finally triggered enough concern that lawmakers passed some environmental legislation, including the Clean Air Act.

As a result of this and other efforts, pollution levels in California have dropped in recent decades. Despite improvement, the 2006 and 2007 annual reports of the American Lung Association (ALA) ranked Los Angeles as the most polluted city in the country, for both short-term and year-round particle pollution. In 2008 the ALA ranked Los Angeles the second most-polluted, and the city again had the highest year-round particulate pollution. In addition, Los Angeles groundwater is increasingly threatened by Methyl Tertiary Butyl Ether (MTBE) from gas stations and perchlorate from rocket fuel (South Coast Air Quality Management District 2010).

The Greater Los Angeles area is the uncontested monument to the carbon-heavy, fossil fuel–driven 2IR. It represents everything that was once glorious about cheap fossil fuel, which now threatens the world with global warming and environmental degradation. Neither Los Angeles nor its lifestyle will easily adapt to the GIR. However, for the sake of the planet, Los Angeles will have to make significant changes, and the sooner the better, if we are to avoid a catastrophic increase in global temperatures.

The answer for Los Angeles, other world cities, and all our communities, is sustainability. Cities must become more walkable, bike friendly, and livable. They need to focus on environmental sustainability as well as economic sustainability. The quality of urban space must improve. The architecture should be inventive with sensitive urban design and a dynamic atmosphere. Communities must promote sustainable living and sustainable business development, along with the required infrastructure needs of water, recycling, transportation, waste, and materials. Above all else, a sustainable community needs to generate renewable energy and use energy storage and a smart grid system to balance and smooth out intermittent power demand and supply.

Sustainable communities started in Europe and Japan as a reaction to the Arab oil embargo of the mid-1970s. They gained a toehold in Germany, then Denmark and Holland. These countries had historically used wind power as a source of renewable energy, and so they quickly embraced sustainability. They were followed by other Scandinavian countries that realized that the North Sea oil wasn't going to last.

Today, major wind farms and biomass generation plants dot Britain. Pushed by the EU, Spain, Italy, and the other European countries have adopted policies and governmental programs to move sustainable communities and the GIR forward, reaping environmental and economic benefits along the way.

With a long tradition of conservation and sustainability, Japan led the movement in Asia. After their modernization blitz during the first half of the 20th century ended with defeat in World War II, the Japanese regained their historic sense of sustainability. Of course, their entrance into the GIR was also triggered by a glaring lack of natural energy resources. The Koreans, and later the Chinese, leaped into the GIR and now most of Asia embraces it. Granted, the region has horrendous pollution and environmental issues to overcome, but Asia is rapidly developing renewable energy technologies. China now leads the world in solar and wind technology production (Chan 2011).

The common theme in all these nations, and a theme that is lacking in America, is the notion that sustainability starts at home, in the behaviors and values of families. Communities are usually described as a group of interacting people living in a common location. Psychologists describe a community as one that includes a sense of membership, influence, integration, and fulfillment of needs, as well as shared emotional connections. Generally, communities are organized around common values or beliefs. They share resources, organize around a political structure, agree on preferences, needs, and risks, and agree to tax themselves for the benefit of the whole.

A sustainable community has these elements, plus a core value of conservation, a respect for natural resources, and an appreciation for the environment. The concept of "achieving more with less" is broadly endorsed. Sustainability is a community-centric activity: the more focused and integrated the community, the more it has a chance of achieving sustainable development. Along with sustainable development come the benefits of a cleaner environment and a healthier lifestyle.

Besides these core values, sustainable communities combine common social activities with business development and job creation. This concept was first defined as *sustainable development* more than a quarter of a century ago. The Brundtland Commission, convened by the United Nations in 1983, introduced the term in its *Our Common Future* report in 1987. In that same report (named for its chairperson Gro Harlem Brundtland, prime minister of Norway at the time), the UN commission addressed emerging concern about "the accelerating deterioration of the human environment and natural resources and the consequences of that deterioration for economic and social development" (Brundtland 1987).

Even in the 1980s, the UN General Assembly recognized that environmental problems were global in nature. Subsequently, it was in the common interest of all nations to establish policies for sustainable development. The Brundtland Commission defined sustainable development as meeting the "needs of the present without compromising the ability of future generations to meet their own needs" (Brundtland 1987). Today, the term describes how a community's economic concerns interact with its natural resources. Addressing large global problems locally can generate new creative ventures and opportunities, which then provide strong business reasons to pursue sustainable development.

All sustainable communities must address the essential infrastructure elements of energy, transportation, water, waste, and telecommunications (Clark 2010). The critical component is renewable energy power generation. Renewable energy provides power generation in harmony with the environment and economic development (Clark 2010). Europe and Japan have developed numerous communities that are sustainable and secure through the use of their own renewable energy sources, augmented by storage devices and emerging technologies.

On a national level, Denmark is close to a classic sustainable community model. With gasoline costing $10 a gallon, Danes prefer to take public transportation. In the mid-1990s, Denmark established a CO_2 tax to promote energy efficiency, despite discovering offshore oil. Since 1981, Denmark's economy has grown 70 percent and energy consumption is almost flat (Lund and Ostergaard 2009).

Denmark's early focus on solar and wind power has paid off, as those sources now provide more than 16 percent of its energy. One third of the world's terrestrial wind turbines now come from Denmark, and Danish companies Danisco and Novozumes are two of the world's most innovative manufacturers of enzymes for the conversion of biomass to fuel (Lund and Ostergaard 2009). As an additional benefit, these industries have provided green jobs.

On a local or regional level, a sustainable community must have three components:

- First, it must have a master strategic plan for infrastructure that includes renewable energy, transportation, water, waste, and telecommunications, along with the traditional dimensions of research, curricula, outreach, and assessments.
- Second, facility planning must be addressed from a green perspective. There is an array of issues pertaining to the design, architecture, and sitting of buildings that affect sustainability. The community needs to consider efficient orientation and multiple-use design. Developing dense, compact, walkable communities that enable a range of transportation choices reduces energy consumption.
- Third, a sustainable community is a vibrant, "experiential" applied model that should catalyze and stimulate entrepreneurial activities, education, and creative learning, along with research, commercialization, and new businesses.

As the sustainable movement has developed, communities have sought out policies that direct facilities and infrastructures to be green. Originally, there were several certification processes; however, Energy Star and the U.S. Green Building Council's Leadership in Energy and Environmental Design (LEED) certifications are becoming universal. There are regional LEED certification councils in most parts of the world, and there are now community neighborhoods or clusters of buildings using LEED standards. This set of criteria reflects broader concerns for clusters of buildings with integrated designs for basic infrastructure needs.

A key component to buildings today is their design (LEED standards) so that they are environmentally sound. The design and construction of buildings and clusters must be addressed as the shift continues to more decentralization and onsite power through renewable energy production.

Small, relatively self-contained communities or villages within larger cites and regions are more easily made sustainable. These communities must reduce or eliminate their dependency on central grid energy sources that come from fossil fuels such as coal, natural gas, and nuclear power.

Local energy generation sites are known as onsite power generation systems. In Europe, these systems usually combine heat and power. Most integrate distributed power from a central grid, but when the systems are based on renewable energy sources, they become sustainable. For a community to be sustainable, it must include renewable energy resources such as wind, solar, biomass, ocean, geothermal, and "run of the river" (not large hydroelectric dams or nuclear energy plants) energy sources that do not negatively impact the environment.

Local onsite power can be more efficiently used and can use the region's renewable energy resources. For example, Denmark's many sustainable communities are generating power with wind and biomass, combined to provide base-load energy power. Denmark has a goal of 50 percent renewable energy generation (primarily from onsite and local resources) by 2020 (Lund and Ostergaard 2009). The country is well on its way to meeting and perhaps exceeding that national goal.

Local communities that want to be sustainable now use onsite renewable power sources for clusters of buildings, like those used for colleges, local governments, residential housing developments, factories, storage and factory facilities, shopping malls, and office complexes.

Distributed or onsite energy systems on local and regional levels provide an important ingredient for sustainability. Europe's combined heat and power systems were developed to meet local needs, reduce the use of fossil fuels, and help communities become energy independent and more self-sufficient. Some U.S. communities are now developing similar systems focused, in many cases, on co-generation or combined cycle (the combined production of heat and electricity) arrangements using renewable energy resources.

Agile Sustainable Communities

The 2IR energy model was to generate power from a central plant with fossil fuels or nuclear energy, and then transmit it over a grid to communities located great distances from the power source. The GIR model is to generate renewable power near the end-users. During this transition, a hybrid model is developing. This new model is agile because it can accommodate both green onsite power generation and grid-connected power. Agile systems combine local onsite renewable energy power with traditional power central grids hundreds of miles away, and manage them both to meet local system power needs.

The new agile local onsite power system is efficient, smart, and rooted in renewable energy power generation. Although there is still a central power grid that usually depends heavily on fossil fuels to generate power, agile systems allow and even encourage onsite power from renewable energy sources.

Agile systems generate power from renewable energy sources such as solar, wind, water, geothermal, and biomass, and then disperse the energy on either a large or small scale. These distributed energy systems can be formed and operated on the local level to serve targeted communities.

College campuses are perfect examples where a central grid can be combined with green buildings that have rooftop solar panels or in some cases wind or biomass for local renewable energy generation. Solar or wind power can generate significant amounts of power for clusters of buildings, like those at colleges or shopping malls, but on days that are not sunny or windy, and at night, these buildings must rely on the central energy grid.

Sustainable and smart agile communities represent a new community model. Historically nations, states, and cities wanted to control and centralize power and authority. The 2IR central grids required long transmission lines, pipelines, or shipping to deliver energy. The standard approach was for municipalities to manage the capital costs for the actual central plant with its processing of raw materials but for ratepayers to absorb the transmission costs.

Regional and community-level solutions are now needed to address the challenges of global warming and climate change. Rather than having centralized power plants that use fossil fuels or nuclear power to generate energy and then transmit it over power lines, local onsite generation of power from renewable sources is better for the environment and far less expensive.

The energy needs for communities around the world are growing more complex as population increases, cities expand, and power demands climb. Air and water pollution cause serious health problems for young and old. Local and national governments are now implementing carbon dioxide regulations to stop pollution. Meeting the challenges of supplying energy for increasing demand, while reducing carbon emissions, calls for more complex and creative solutions. It requires energy efficiency, renewable energy generation, and new systems to change the way people live and think about using electricity.

In the United States, California is on the forefront of this transition. Several small California communities are experimenting with producing their own energy from renewable sources. They are models for other communities that want to generate their own power from renewable sources. While some communities may be better-suited for solar, or PV, systems, others (mountain communities, for example) may be able to use river water flow to generate power. California has considerable amounts of geothermal power as well as significant numbers of wind farms and solar installations. Hydrogen can be produced from renewable or green sources and then stored close to the needs and demands of communities.

Consider how energy systems are evolving today. Agile energy systems are a combination of local or regional systems hooked to central grids. In the future, the central grid will be used for redundancy and back-up purposes or act as a battery for energy storage when the sun is not shining and the wind is not blowing. These agile systems are not just technologies or market mechanisms, but a new model that is part of the GIR.

Sustainability Starts at Home

Researchers and political decision makers around the world are slowly recognizing that they need to do something about climate change and global warming. Sustainability is achievable. It can be done, and must be done, at the community level. Block by block, city by city, region by region, communities can change how they live.

There is nothing new about either electric or hybrid cars. Germany's Lohner-Porsche Carriage originally developed hybrid electric cars in 1903 (Clark and Vare 2009). Until Toyota's Prius, hybrid vehicles were not commercially available, due to economic and political pressures from the oil and car companies favoring fossil fuels and internal combustion engine power sources. The American muscle car culture was highly profitable, and it dies a hard death.

Hybrid cars are not, by themselves, enough to achieve sustainability. However, the use of hybrid technologies provides a short transition period (from 5 to 10 years) until the next generation of all electric, hydrogen, and other environmentally sound vehicles are in the market place. Energy consumption must also be reduced by developing dense, compact, walkable communities that enable a range of transportation choices.

Communities throughout the world are entering the GIR and developing plans for sustainable futures. The use of old 2IR energy sources—fossil fuels and nuclear power—is losing political and community support as advanced GIR technologies are developed. Britain, a country that for years maintained its prosperous lifestyle as a financial go-between for the Middle Eastern oil trade, recently hit a milestone of 1gW of installed offshore wind turbine capacity in April 2010 with the completion of the Gunfleet Sands and Robin Rigg wind farms. Future plans call for the development of 25gW from offshore wind farms, with more than 7,000 wind turbines (Gipe 2011).

In South America, Brazil is 95 percent energy independent through a combination of sugar cane ethanol and domestic oil supplies (International Energy Agency 2010). To the west, Chile is developing renewable energy as a power source, after numerous public demonstrations against more hydroelectric dams. Chile is one of the world's most beautiful countries, with wild and pristine regions interlaced with extraordinary free-flowing rivers. The Chilean public is adamant about keeping the rivers free from additional hydroelectric dams, and the government is responding by opening the door to renewable energy. In Mejillones in the Atacama Desert region of northern Chile, Algae Fuels S.A. consortium is using microalgae in second-generation biodiesel production. Wind and solar energy development is also headed to the region.

In Denmark, Frederikshavn is incorporating multiple renewable energy technologies strategically over time to make the entire city energy independent. By 2015, the city plans to be 100 percent energy independent by using renewable energy systems like wind and biomass. By 2010, Frederikshavn had already achieved 45 percent of its goal (Lund and Ostergaard 2009).

In other parts of the globe, China and Spain are developing sustainability through public policies that support renewable energy power generation, such as a FiT, which fix rates to provide rebates to consumers. Communities in Japan have been sustainable for many years, since Japan must either import all of its energy or generate it within the island nation. Increasing,

Japan is using renewable energy, and until recently, it was the world leader in solar manufacturing companies. Now Italy has been active regionally, for different reasons, but primarily due to the national historical regionalization and its city-focused policies and programs. In the Baltic Sea region that was part of the Soviet Union, Lithuania has been active and begun a national focus on sustainability.

The United States has not been quick to join this movement, but several American city governments, such as San Francisco and Santa Monica in California, are supporting LEED-designed buildings, and encouraging local renewable energy generation including solar, wind, and ocean power, along with electric vehicles and hydrogen-fueled cars.

A Most Livable City

Consider Melbourne, Australia, a city of nearly four million people that has had a drought since 1997. The reality of this, and other significant environmental challenges, prompted the city to start a campaign of sustainability despite its large urban footprint, 2IR sprawl, and the demand for infrastructure and services. Over the years, the city has made extraordinary strides. Melbourne was acknowledged in 2010 by *The Economist* as one of the World's Most Livable Cities ("Livability Rankings" 2010). The city has launched a program to make it the world's most sustainable community by 2020.

How did Melbourne, a city once sprawling and auto-centric, manage to turn itself around? Low rainfall, and then the extended drought, coupled with high temperatures, presented the first big challenge. In response to dropping water supplies, the city implemented water restrictions and a range of other options, including water recycling schemes, incentives for household water tanks, gray water systems, water consumption awareness initiatives, and other water saving and reuse initiatives.

The Australian government considered other approaches, as well. It proposed developing a desalination plant on the southeast coast of Victoria (the state in which Melbourne is located), but numerous independent reports found that sustainable water management would be the best solution ("Livability Rankings" 2010). In 2002 Melbourne responded to climate change with a plan to reduce carbon emissions to net zero by 2020. Then the city began a serious effort to move forward to sustainability.

In 2005 the Victorian government developed the Melbourne 2030 Plan, a strategy for sustainable growth to 2030. During that time, Melbourne's population is expected to increase by another one million people, and

Melbourne 2030 is a plan to manage that growth and development. The government also wants to maintain the city's attraction as a livable and prosperous place for residents, businesses, and visitors.

The main thrust of Melbourne 2030 is to protect the livability of established areas and to accommodate major change and increased housing density in underdeveloped areas and urban centers (activity centers) that have room for growth.

Achieving sustainability requires an integrated approach to decision making. This means taking a long-term view while ensuring that economic, social, and environmental implications are considered. Australia's National Strategy for Ecologically Sustainable Development provided a framework for achieving long-term sustainability whose core objectives are to:

- Safeguard the welfare of future generations
- Improve equity within and between generations
- Protect biological diversity and maintain systems essential to support life

A key principle is that, where there are threats of serious or irreversible damage, lack of full scientific certainty shall not be used as a reason for postponing cost-effective measures to prevent the threat from being realized.

As a follow-up to Melbourne 2030, the Melbourne City Council initiated *Future Melbourne,* the most ambitious community consultation project it had ever undertaken. The council aims to make Melbourne one of the world's most sustainable cities.

Future Melbourne is a long-term plan for the future direction of all aspects of city life. Developed by the community, the draft plan aims to build on Melbourne's livability. It sets out goals for the future, identifies key trends and challenges, and outlines strategic growth areas for the city in four specific strategic areas for being sustainable.

Adapting for climate change

A major part of the eco-city plan is adapting while still trying to mitigate or stop climate change locally. The plan emphasizes the importance of better understanding the city's dependence on resource-based industries, the capacity of its infrastructure and the needs of the population.

The management of water shortages and flooding will be improved through storm water harvesting. There are also plans to ensure that buildings and infrastructure can mitigate the impacts of climate change. In addition, the city will use cooling from parks and rooftop gardens to deliver eco-system cooling services as well as conduct passive shading to reduce the impact of heat stress.

A resource efficient city

Melbourne plans to become a resource and material-efficient city, through partnerships with various private and public groups and organizations,

The council promotes the use of backyard and rooftop vegetable gardens as well as encourages greater waste reduction, re-use and recycling. They also incorporate waste management facilities to assist the public to reduce waste sent to landfill.

To achieve resource efficiency, Melbourne plans to:

- Reduce household waste in the city.
- Reduce commercial waste in the municipality.
- Develop a municipal ecological footprint and target.
- Develop and implement a more economic and efficient waste collection and processing system.

A dense urban center

Due to the city's low density, the council identified a need to focus on sustainable modes of transportation, such as walking and cycling. Moreover, new urban growth should be focused around existing and proposed rail stations and tram routes. The production, distribution, and consumption of locally and regionally grown food are other ways to improve the city's sustainability. The number of people living in the city will also be increased from 42 to 65 per cent.

The city as a catchment

The "city as a catchment" philosophy helps to determine the flow and amount of water moving through the municipality and the pollutants that are carried with these flows. The program aims to reduce residential and office water consumption as well as the pollution entering the city's waterways.

As part of its sustainability, Melbourne is intent on replacing its use of brown coal, which is the dirtiest energy source on the planet, with energy from renewable sources. A plan proposed by the Greens, a major political force in Melbourne politics, calls for the city to make a complete switch to renewable energy. Called Repower Melbourne, the 2010 proposal sets out a suite of policies that will drive the transformation to clean energy. The proposal calls for neighborhood "solar farms," and suggests covering the roofs of public buildings with solar panels.

Despite its large urban footprint and 2IR sprawl, Melbourne is creating a sustainable community that is interconnected and accessible. Over the years, Melbourne has invested heavily in inner-city renewal, parks, and green wedges. It has created a vibrant, walkable city center and has taken bold steps to achieve significant improvements in urban life. There are countless websites urging residents to adopt green practices, recycle, conserve water, and contribute to the city's sustainability drive. Vegetable gardens spill over onto the sidewalks, and the residents' excitement is creating a vibrant and dynamic atmosphere.

If Melbourne, a large, sprawling city in a sensitive, drought-constricted environment, can set out on a path to become one of the world's most sustainable cities, then there is hope for other large cities trapped in the 2IR model. Who knows, maybe the residents of North America's mega-highway-laced cities can turn off the carbon highway and move toward a greener future.

Schools, Colleges, Universities, and Other Clusters of Buildings Are Sustainable Communities

Some of America's more progressive educational institutions are inching toward sustainability. In California, solar panels are starting to dot the rooftops and cover the parking lots of many schools and colleges. Many colleges recycle waste as part of school activities, and consider energy efficiency when planning and upgrading facilities. In some schools, energy efficiency and water conservation are included in the curriculum.

Even in Los Angeles, the quintessential 2IR dream world, where concrete freeways stretch for hundreds of miles and the internal combustion engine is a way of life, colleges are becoming sustainable. In fact, the Los Angeles Community College District (LACCD) is a key educational system when it comes to sustainability. LACCD is largest community college district in the United States, serving more than 250,000 students annually at nine colleges spread throughout 36 cities in the Greater Los Angeles area. Each college aggressively recycles trash and waste, conserves water, and saves energy. In 2001 and 2002, the LACCD met high U.S. Green Building Council standards for the buildings on each campus and led a national movement toward sustainable academic campuses on all levels.

Community support has made the difference. Los Angeles County voters passed two construction bonds worth more than $5.3 billion to fund the massive transformation of its nine college campuses. The LACCD created a sustainable community vision that called for energy efficient buildings and renewable energy systems for each campus. While there has been some concern and criticism of the bond expenditures, the LACCD retrofitted buildings to conserve electricity, natural gas, and water, as well as to promote recycling. It installed solar panels on rooftops and over parking lots, and augmented them with wind turbines. It also installed biomass fuel systems coupled to storage devices to provide continuous energy.

As part of the greening effort, the LACCD has pushed to have its buildings achieve gold (or higher) levels of LEED certification, while making its nine campuses energy independent and carbon neutral. The district was on track to have 90 buildings LEED-certified by 2012.

The LACCD has brought this sustainability effort to promote environmental awareness in the transfer curriculums, in job training programs, and in the daily practice on all nine campuses. A Sustainability Committee was created that has met monthly since 2007, and more than 90 faculty, administrators, and staff members are involved.

The scope of the LACCD program is truly inspirational, despite debate over costs and allocation of resources that promoted dramatic leadership changes in 2011. Already other colleges and universities have developed curriculum and training programs for their students who are interested in new jobs in the emerging green industry. A national Association of Sustainable Higher Education was formed and educational and curriculum changes have been made to help develop workers that can create, operate, and maintain sustainable communities. Slowly, the movement toward sustainable communities has become part of America's institutions of higher learning. This has also moved the nation's architecture, engineering, and construction professions toward sustainability, while expanding the scope of more traditional disciplines like economics, law, health and medical care. In the end, the prime lesson learned is to have objective due diligence done from the very beginning that is overseen by an informed citizens organization that is mandated by law, as in California's Proposition 39, which requires such accountability (Clark 2010).

Hotels and Resorts Are Sustainable Communities

In a combination of peer pressure and the fight for survival, the world's luxury hotel industry has started to compete as green and sustainable hotels. Large luxury hotels are extraordinarily attractive environments, full of lavish furnishings, uniformed attendants, wonderful food, and dramatic interiors. In reality, they are mini-communities, providing the standard infrastructure elements of energy, transportation, water, waste handling, and telecommunications for a small, transient, and demanding client population. Hotels are energy hogs, constantly using energy and resources, even at night and during "down" times. Luxury hotels have amazingly complex inner workings that include kitchens, freight elevators, laundry rooms, parking structures, and enormous HVAC (heat, ventilation, and air conditioning) boilers, all of which wreak havoc on the environment.

Hotels require constant food preparation and room servicing, with daily washing of sheets, towels, and tablecloths. They maintain a roaring stream of tiny soaps, bottles of shampoo, and other personal care items, and leave lights, and either air conditioning or heat, on continuously. For these reasons and more, hotels gobble up water and energy, while generating massive amounts of paper, food, metal, plastic, aluminum, and glass waste.

Luxury hotels are the worst offenders. At hundreds, sometimes thousands, of dollars a night for a room, a luxury hotel cannot trim back on customer service. The luxury hotel industry knows their clients are particular; a major facility will spend $1 million a year just to keep furnishings crisp and current. For this reason, luxury hotels have been particularly resistant to change, due to fear that once word gets out that a hotel is cutting back, it will lose its clientele.

The worldwide economic downturn (more than concern for the environment) has forced the hotel industry to rethink its operational costs and reevaluate its extravagant use of resources. Resources are expensive and the hotel industry is turning to energy efficiency and green activities—like ozone laundry, retrocommissioning, and LED (light emitting diode) lighting—to hold costs down. A hotel aiming for sustainability uses resources wisely, and conserves and preserves by saving water, reducing energy use, and cutting down on solid waste. Incorporating these waste reduction techniques into hotel operations is a win-win for the environment and the hotel. Not only will greener hotels save resources and reduce pollution, they will also cut down on operating costs while increasing profit margins.

This extraordinary progress toward the GIR probably started with the eco-friendly hotels in countries with fragile environments, like Costa Rica and Australia. However, the big luxury hotels of Mumbai, Macau, and Dubai are taking sustainability seriously, and are reaching out to architects and engineering firms for advice on LEED certification and renewable energy generation.

In the fiercely competitive luxury hotel market in San Francisco, sustainability has become a marketing tool. In this tourist-driven market, the luxury hotels are lining up to promote their "greenness." Market studies are now convincing hotel owners that, given the choice, high-end clients prefer a hotel that is environmentally responsible.

In a tight, hyper-competitive environment, differentiation, as well as holding down costs, is the key to success. Despite the initial capital investment required for energy efficiency retrofits, hotels are turning toward sustainable practices. Even the Grand Hyatt, one of San Francisco's oldest and most conservative luxury hotels, is planning a massive remodeling project, with a "living wall" as a design focal point in the main lobby. The hotel also plans to put solar panels on the roof. The GIR luxury hotel movement has begun in San Francisco, and eventually, it will spread throughout the industry.

Creating and Implementing Sustainable Communities

Creating sustainable communities is an extraordinarily complex task. It begins with addressing key infrastructure elements—energy, transportation, water, waste, and telecommunications—and extends through incorporating the belief systems, values, and behaviors of residents. Codes and standards are required to guide how buildings are designed, sited, and constructed. Certification programs like LEED provide guidelines, expertise, and political influence on how to construct and retrofit environmentally sensitive facilities that allow for the maximum use with fewer resources. Public policies establishing goals are needed to reduce GHG emissions and set thresholds and benchmarks for renewable energy power generation.

Cities can follow the example set by Melbourne. The city held multiple community meetings on sustainability to educate residents. Then they developed plans that could be publically reviewed, and created websites that allowed residents to keep tab on which programs were being developed and the progress that was being made. To encourage participation, they also conducted campaigns to create awareness and encourage sustainable practices by the residents ("Livability Rankings" 2010).

Emerging technologies are providing additional tools for achieving sustainability. This is not only apparent in the development of new renewable energy technologies, but also in the discovery of new ways to conserve valuable resources, particularly water. There are devices just coming to market that will help minimize the water used in large HVAC systems. These retrofit systems are gaining credibility in water-constrained locations such as California. They reduce the amount of water used in industrial scale boiler and chiller systems. Not only do these systems conserve the amount of water required to run the equipment, but they greatly reduce the outflows. A city like San Francisco not only has high water costs, but the sewer costs are equally high. One recently installed device on a large San Francisco hotel is projected to save $50,000 a year in sewer costs alone. This savings, combined with the reduction in the original water use, means that the cost of the device is recovered in just a few months (Dugar, Ting, and Partin 2009). Ozone laundry conversions for commercial laundry systems provide similar benefits and a comparable short-term payback.

Transportation is a critical need in today's frenetic world, but the idea of compact, dense, walkable, and bike-friendly cities conflicts with the dream of personal transportation. Unfortunately for global warming, one of the biggest desires people in a developing nation have is the possession of a personal transportation device such as a car.

Automobiles are central to the rising middle-class lifestyle and a key component of globalization and modernity. Unfortunately, cars and trucks powered by fossil fueled internal combustion engines are primary contributors to carbon emissions, GHGs, and global warming. Our love for speed, loud exhaust pipes, and Cordoba leather bucket seats is pushing us ever closer to a hot, dry planet with too much sunlight, too many powerful hurricanes, and not enough food and water. If we don't find a new approach to transportation, everyone will suffer.

Dense, walkable communities would help, and simple, cheap and plentiful public transportation would go a long way toward solving the problem. However, cars are what people want. One reasonable solution comes from Toyota, the world's biggest hybrid automaker. Hybrid and electric cars aren't particularly new, but it took the 1973 global energy crisis to give the world the Toyota Prius.

The Arab oil embargo starkly reminded the Japanese of their historic dependence on foreign oil and gasoline. This was the same reality that drove them into World War II, but this time the Japanese decided to seek environmentally friendly technologies to power buildings and vehicles. This was how the modern hybrid car came to be in the early 1990s in Japan. The government knew that greater fuel efficiency, along with higher mileage per gallon, would mean less use of fossil fuels and so they encouraged Toyota to develop such a car. The Prius was born.

Japan's creation of the world's first popular hybrid car was a great irony for America. The regenerative braking technology, which is the foundation for the Prius and other hybrids, was American. This inspired technology was created at California's Lawrence Livermore National Laboratory in the mid-1990s and offered to the American automakers. When they rejected it, Toyota licensed the technology (Clark and Vare 2009).

The Prius went on sale in Japan in 1997, where it became the first mass-produced hybrid vehicle. It was introduced worldwide in 2001, eventually gaining acceptance in Southern California, where driving is part of survival and the cost of fuel is high. Southern California also has a growing interest in fighting smog and protecting the dry environment. Southern California's climate and environment are key to attracting tourists and fostering businesses where employees can enjoy the weather and scenery ranging from beaches to valleys to mountains and deserts.

Today, the Prius is the most fuel-efficient gasoline-using car sold in the United States, according to the U.S. Environmental Protection Agency (EPA). The EPA and the California Air Resources Board (CARB) also rate the Prius as among the cleanest vehicles, based on smog-forming and toxic emissions. The Prius is sold in more than 70 countries and regions, with its largest markets being those of Japan and North America. In May 2008, global cumulative Prius sales reached the milestone of one million vehicles, and by September 2010, worldwide cumulative sales reached two million. The United States is the largest market, with over a million vehicles sold (Toyota, http://www.toyotausa.com/2010).

The world's other automakers did not fully realize the consumer demand for environmentally sound personal transportation. Toyota's Prius and Camry, and now Honda's Civic and Clarity are examples of the automobile's move into the GIR. The Prius, for example, with its combination of electric and gasoline motors, provides a constant 45 miles per gallon, steady reliability and adequate performance.

Hybrid vehicles are just the beginning. Japan has taken the next step and developed small-scale hydrogen fuel cells to power small buildings. Currently, there are just over 2,000 homes powered by these small stations, with the goal that by 2020, a quarter of all the homes in Japan will be powered by fuel cells (Funaki and Adams 2009). Currently, hydrogen for these fuel cells is produced from natural gas; however, soon, renewable energy electrolyzers will take over.

To develop sustainable communities, the world needed a plan or a set of concepts that would help lead to that goal. Melbourne set the ball rolling by establishing a set of principles to guide the city toward sustainability. In 2002 the Local Government Session of the Earth Summit in Johannesburg adopted the Melbourne Principles ("Livability Rankings" 2010). The principles consisted of 10 short statements on how cities can become more sustainable. They are designed to be read by decision makers, and provide a starting point on the journey toward sustainability.

The Melbourne Principles provide a conceptual framework for moving today's sprawling, ever-growing cities toward sustainability. In addition, agile, sustainable communities must develop strategic plans for sustainable processes for energy, waste, water, transportation, and telecommunications. As well, they must develop the financial resources to implement them. These infrastructure elements are integrated; transportation and energy are connected because transportation systems should use renewable energy. The same is true for water pumping and surface transportation, particularly in places like California that require enormous amounts of energy to move large quantities of water across miles of surface area.

Each sustainable community must retrofit the traditional central power plant into one using renewable energy generation and smart grid distribution. Further, the sustainable infrastructure systems must provide for recycling, waste control, water and land use, as well as green building standards that require energy efficient and compact housing. Today, agile sustainable and smart communities are necessary to reduce environmental pollution and to provide a green world for tomorrow. The solutions to global warming and climate change exist now; we need to design, finance, and implement them.

Chapter 7

Renewable Energy Integrated Systems

America's fossil-fuel addiction results in a huge and disproportionate amount of energy consumption, more than any other nation in the world. In fact, the United States consumes about 20 percent of the world's energy each year. According to the U.S. Department of Energy, Energy Information Administration's 2011 "Annual Energy Outlook," America will consume more than 100 quadrillion BTUs (a measure of thermal energy) in 2011. This averages to about 14,000 kilowatt hours (kWh) per person. As the U.S. economy becomes more technical and knowledge-based, energy consumption is expected to grow about 20 percent, to 4,880 billion kWh by 2035 (U.S. Department of Energy, Energy Information Administration 2011).

Most consumed energy is in the form of electricity (about 85%). Of the remainder, about 12 percent is from natural gas, and the rest is from steam (U.S. Department of Energy, Energy Information Administration 2011). Electricity is generated by electromechanical generators, primarily those driven by heat engines fueled by chemical combustion or nuclear fission, or by kinetic energy from sources such as flowing water.

The basic method to create electricity was discovered by Michael Faraday, a British scientist around 1830. Faraday figured out that electricity could be generated by moving a loop of wire, or copper disc, between the poles of a magnet.

The United States has been using electricity as a source of energy since the 1880s, when several inventions made widespread use of electrical power possible. Power transformers provided a way to generate electrical power from a centralized location. Alternating current power lines provided a way to transport electricity at very low costs across great distances by raising and lowering the voltage. These developments paved the way for America's first power plants, which ran on coal or water power.

Coal is the main source of power for generating electricity in the United States, but other emission polluting sources are increasing in popularity, such as natural gas. According to the U.S. Department of Energy, Energy Information Administration's 2011 "Annual Energy Outlook," about 85 percent of U.S. electricity is generated from fossil fuels, with about half coming from coal, due to government tax and finance incentives continuing since the 1880s. Over the past several years, the capacity of natural gas plants has been increasing, and now about 23 percent of U.S. electricity is generated from natural gas. That and shale oil (from dirt) are promoted by the fossil fuel industry as the answers to America becoming energy independent, since so much exists within the United States and nearby Canada. The permanent damage to the environment (land use, factoring, and sequestration, etc.), increased atmospheric emissions, along with transportation and security costs, are never discussed or calculated. Another 20 percent of the nation's electricity is generated by nuclear power and 6 percent by large-scale hydroelectric facilities. Wind and solar energy sources contribute about 1 percent (U.S. Department of Energy, Energy Information Administration 2011).

Electricity is used primarily for lighting and for making buildings livable and comfortable. Heating and air conditioning use the bulk of the nation's energy, about 40 percent (U.S. Department of Energy, Energy Information Administration 2011). Data centers, the keystone to a knowledge-based economy, are another major user of the nation's energy.

Coal's dominant role in generating America's electricity is due to its domestic abundance and seemingly inexpensive extraction. Unfortunately, very few economists include the cost of the environmental damage caused by using coal to generate massive amounts of electricity. While coal-generated electricity may be cheap to some consumers because of government subsidies and incentives, the damage that acid rain causes to the environment and human health is painfully expensive to remediate and never calculated. And nobody wants to spend money to offset the millions of tons of greenhouse gases (GHGs) that burning coal releases into the atmosphere. The cost of global warming is a hard concept to understand for people accustomed to cheap electricity.

Renewable Energy—The Key to a Green Industrial Revolution

The GIR is all about renewable, environmentally friendly energy generation, transmission, and distribution. A degrading environment, global warming, severely damaging climate changes, dwindling fossil fuel supplies, and the need to end the continued dependence by the Western democracies on the socially volatile and politically unstable Middle East will force the adoption of alternative and renewable energy sources.

Renewable energy is a source of energy that is not carbon-based and will not diminish. For example, the sun is shining during the day and the wind blows fairly constantly. Both could be used to generate energy, but require some form of storage or feedback technology for when the wind is not blowing and the sun is not shining. For this reason, these forms of energy generation are called "intermittent" and need storage capabilities and integrated systems to provide "base load" or round-the-clock power generation.

The most common renewable energy sources are systems that make use of the wind, sun, water, or a digestive process that changes waste into biomass or recycles waste for fuel generation. Other renewable sources include geothermal, "run of the river," and now increasingly, bacteria and algae.

Wind Power Generation

Wind has been used as power source for tens of thousands of years. The ancients used wind power for sailboats, and this original technology had a major impact on the first windmills, which used sail-like panels to catch the wind.

The first documented use of windmills was in Persia around A.D. 500 to simplify tasks like grain grinding and water pumping (Dodge 2009). The designs used vertical sails made of reed bundles attached to a central vertical shaft by horizontal struts. The Chinese statesman Yehlu Chhu-Tshai constructed a windmill of similar design in China in 1219 (Dodge 2009). One of the earliest and most scenic applications of wind power is the extensive use of water pumping machines on the island of Crete. Today, hundreds of these ancient sail-rotor windmills still pump water for crops and livestock.

The Europeans developed windmills, or "post mills," around 1300. The earliest illustrations (dating from around 1270) show a four-blade mill mounted on a central post. European mills used wooden cog-and-ring gears to translate the motion of the horizontal shaft to vertical movement to turn a grindstone. This gear was apparently adapted for use on post mills from the horizontal-axis water wheel developed by Vitruvius, a Roman writer, architect, and engineer active in the first century B.C.

The Dutch refined the design and created the tower mill around 1390. They affixed the standard post mill to the top of a multi-story tower, with separate floors devoted to grinding grain, removing chaff, storing grain, and (on the bottom) living quarters for the windsmith and his family. Both the post mill and the later tower mill had to be oriented into the wind manually, by pushing a large lever at the back of the mill. The windsmith's job included not just optimizing windmill energy and power output, but also protecting the mill from damage by furling the rotor sails during storms (Dodge 2009).

It has taken 500 years, and countless incremental improvements in efficiency, to perfect the windmill sail. By the time the process was complete, windmill sails had all the major features recognized by modern designers as being crucial to the performance of modern wind turbine blades. These mills were the power motors of preindustrial Europe.

For hundreds of years, the most important application of windmills at the subsistence level has been pumping water using relatively small systems with rotor diameters of one to several meters. These systems were perfected in the United States during the 19th century, beginning with the Halladay windmill in 1854.

The most important refinement of the American fan-type windmill was the development of steel blades in 1870. Steel blades could be made lighter and worked into more efficient shapes. They worked so well, in fact, that their high speed required a reduction (slow-down) gear to turn the standard reciprocal pumps at the required speed (Dodge 2009).

Wind turbines translate the kinetic energy of moving air into mechanical energy, and then transmit that energy into a generator that produces electrical power. Large-scale turbines have a horizontal-axis design. Two or three rotor blades are mounted atop a tower, similar to the blades of an airplane propeller. The movement of wind across the blades generates lift, spinning the shaft connected to an electric generator. The output is a function of wind speed and the size of the turbines rotors.

The first use of a large windmill to generate electricity was a system built in Cleveland, Ohio, in 1888 by Charles F. Brush. In 1891 Poul La Cour, a Danish scientist, developed the first electrical output wind machine to incorporate the aerodynamic design principles (low-solidity, four-bladed rotors incorporating primitive airfoil shapes) used in the best European tower mills. By the close of World War I, the use of 25kW electrical output machines had spread throughout Denmark, but cheaper and larger fossil-fuel steam plants soon put the operators of these mills out of business.

A utility-scale wind energy conversion system was first tried in 1931 in Russia. Experimental wind plants were also tried in the United States, Denmark, France, Germany, and Great Britain during the period 1935–1970. These demonstration sites showed that large-scale wind turbines would work, but did not lead to a practical, large, electrical wind turbine. European developments continued after World War II, when temporary shortages of fossil fuels led to higher energy costs (Dodge 2009).

In the United States the federal government's involvement in wind energy research and development began in earnest after the Arab Oil Embargo of 1973. Despite the speed with which it was initiated, the program was doomed by political factors and the withdrawal of financial support before success could be achieved. However, by the mid-1980s, the commercial wind turbine market had evolved from the need for small (1 to 25kW) machines for domestic and agricultural applications to intermediate-size (50 to 600kW) machines for utility interconnected wind farms.

California installed the majority of wind turbines until the early 1990s. In California more than 17,000 turbines, ranging in output from 20 to 350kW, were installed in wind farms. At the height of development, these turbines had a collected rating of over 1,700MW and produced more than 3 million MW hours of electricity, enough (at peak output) to power a city of 300,000.

The U.S. wind farm market lagged and gradually declined during the 1980s and into the 1990s. The downturn was triggered by the end of the federal energy credits in 1984 and the phase-out of the California state credits.

This was not the case in other parts of the world. Fear that the Arab Oil Embargo posed a serious threat to their national independence led the nations of northern Europe and Asia to develop policies to encourage wind energy. Wind turbine installations increased steadily through the 1980s and 1990s in these regions. The higher cost of electricity and excellent wind resources in northern Europe created a small, but stable, market for single, cooperative-owned wind turbines and small clusters of machines.

After 1990, most market activity shifted to Europe and Asia. Driven by high utility power rates, cooperatives and private landowners in the Netherlands, Denmark, and Germany installed first 50kW, and then 100kW, 200kW, 500kW, and finally 1.5MW wind turbines. This impressive growth in installations now amounts to over 10,000MW of European wind capacity and has helped support a thriving private wind turbine development and manufacturing industry (Dodge 2009).

In the United States, wind energy development resumed around the turn of the century, buoyed by "green power" initiatives in Colorado and Texas. A variety of new wind projects were installed in Texas, Northern Colorado, the upper Midwest, and in California. The United States is rich in wind resources, and wind power generation has grown slowly, but steadily, in the 21st century.

According to the American Wind Energy Association, today wind power accounts for about 2.3 percent of the electricity generated in the United States, with an installed capacity of over 40,000MW. Wind energy now produces enough electricity to power the equivalent of nearly nine million U.S. homes. Use of wind, instead of fossil fuels, to generate power avoids 57 million tons of carbon emissions each year and reduces expected carbon emissions from the electricity sector by 2.5 percent. Ironically, Texas, which is the world's icon for fossil fuel, is the state with the largest wind capacity (10,085MW). It is followed by Iowa (3,675MW) and California (3,177MW). The Roscoe Wind Farm in Texas (with 780MW of wind capacity) is the world's largest wind farm. With these new installations, the United States is on track to generate 20 percent of the nation's electricity from wind energy by 2030 (American Wind Energy Association 2011).

In fall of 2010, England opened the world's largest offshore wind turbine farm. Thanet Wind Farm is located off England's southeast coast. It took more than two years to build and will generate 300MW of electricity, or about the amount required to power 200,000 homes. It has 100 turbines (each 377 feet tall), spread over 13.5 square miles. With this new capacity, Britain now generates approximately 5gW of electricity from wind, about enough to power all the homes in Scotland. The British government has said it will support the renewable energy industry to achieve its goal of generating 15 percent of its energy from renewable sources by 2020. It gets about 4 percent of its electricity needs now from wind (O'Toole 2010).

Chile, a long narrow country on the western coast of South America, has announced a major initiative for renewable energy, mostly from wind. Chile has an economy that is heavily dependent on the energy-intensive extraction of mineral resources, but has no fossil fuels. It imports large quantities of natural gas from Bolivia and Argentina, two of its main South American rivals. Its electricity needs have been through the use of hydroelectric projects on its river. However, there is growing public opposition to any additional damming of Chile's pristine rivers, which are a main source of eco-tourism and national pride.

In response, the Chilean government has announced a major effort to generate electricity with NCRE, or nonconventional renewable energy. While lacking in fossil fuel resources, Chile is blessed with extraordinary wind resources, particularly in the northern desert area and along the Pacific Coast. Chile has launched a major campaign to encourage the development of wind turbine farms, with the goal of obtaining 20 percent of the country's electricity from renewable sources by 2020 (Spencer and Sheils 2009).

The use of wind, one of the most ancient power sources, is growing rapidly as the world moves into the GIR. According to the World Wind Energy Association, by June 2010, 196gW of electricity were being generated by wind power, which equated to about 2.5 percent of the world's energy production. Eighty countries are using wind power on a commercial basis. Several had achieved relatively high levels of wind power penetration by 2009: 20 percent of stationary electricity production in Denmark, 14 percent in Ireland and Portugal, 11 percent in Spain, and 8 percent in Germany (World Wind Energy Association 2011).

Wind energy, in contrast to fossil fuels, is plentiful, renewable, widely distributed, clean, and produces no GHG emissions during operation. While there is some criticism of wind farms because of their visual impact, any effects on the environment are generally among the least significant of any power source.

Large-scale wind farms are not the only solution. Today's new technology allows wind turbines to be installed in small communities. Even smaller systems can be placed on rooftops to capture the natural flow of air over buildings.

Advances in turbine construction have increased efficiency to the point that wind energy is quickly becoming the most cost-effective source of electrical power. In fact, a good case can be made for saying that it has already achieved this status. The actual life cycle cost of fossil fuels (a figure that would include costs from mining, extraction, transportation, and use, as well as the costs from environmental and political impacts) is not really known, but it is certainly far more than the current wholesale rates. The eventual depletion of fossil fuel energy sources will entail rapid escalations in price, which will require postponing actual costs that would be unacceptable by present standards. And this doesn't even consider fossil fuels' environmental and political costs, which are mounting every day.

The major technology developments that enabled wind power commercialization have already been made, but there will be more refinements and improvements. Based on the way other technologies have developed, the eventual push to full commercialization and deployment of wind power will happen in a manner that is not imaginable today. There will be some sort of major change in the marketplace that will put several key companies or financial organizations in a position to profit. They will take advantage of public interest, the political and economic climate, and emotional or marketing factors to position wind energy technology as the dominant energy source for the GIR.

Solar—Energy from the Sun

The sun is the Earth's primary energy source. It radiates an enormous amount of power, about 170,000 terawatts (TW), and powers almost all processes, biological or otherwise, that occur on the Earth's surface. There is no shortage of energy from the sun's power.

While the argument could be made that most current energy sources such as wind, biomass, and fossil fuels have a solar genesis, two technologies directly convert electromagnetic energy from the sun into useful energy: solar photovoltaic and solar thermal.

Photovoltaic systems convert light into electricity. Silicon cells capture sunlight, including ultraviolet radiation. The sunlight creates a chemical reaction and excites the electrons in a semiconductor, which then generates a current of electricity. This photovoltaic reaction is at the core of solar panel systems.

Though usually considered a development of the space exploration era, the photovoltaic effect was discovered in 1831 by Alexandre Edmond Becquerel, a French physicist. About five decades later, U.S. scientist Willoughby Smith discovered the photovoltaic effect in selenium and realized that illuminating a junction between selenium and platinum also creates a photovoltaic effect. These two discoveries led to the first selenium solar cell construction in 1877 (PVResources 2011).

A series of discoveries in the 1950s led to the 1958 launch of Vanguard I, the first satellite to use solar cells for onboard power. The solar cell system ran continuously for eight years. Several other satellites with onboard solar cell-generated power followed: Explorer III and Vanguard II, launched by the Americans; and Sputnik III, launched by the Russians.

A Japanese high-tech company, Sharp Corporation, known for its electronic devices and not solar energy systems, developed the first usable photovoltaic module from silicon solar cells in 1963, which set the stage for the global modern solar industry. The big solar companies started forming in the 1970s–1980s in the United States, Japan, and Germany (Clark 1999). ARCO Solar was the first to produce photovoltaic modules, with peak power of over 1MW per year (PVResources 2011).

A PV system includes mechanical and electrical connections, mountings, and a means of regulating or modifying the electrical output. Because the voltage of an individual solar cell is low, the cells are wired in series to create a laminate. The laminate is assembled into a protective weatherproof enclosure, thus making a photovoltaic module. Modules may then be strung together into an array. Electricity generated by PV systems can be used directly as a standalone power source, stored, fed into a large electricity grid, or linked with many domestic electricity generators to feed into a small grid.

Silicon crystal and thin film are the two main categories of PV technologies. Silicon crystal is used more often because of its higher efficiency and greater abundance, although the process of refining silicon is expensive. While thin-film technologies have a lower efficiency, they do have the potential to provide solar power at a lower cost per watt than silicon crystals. A new or third generation of PV designs is developing that features low-cost, low-efficiency materials that may or may not achieve significant scale.

Germany made a huge commitment to solar power at the beginning of the 21st century, a move that surprised the industry and really kick-started the modern solar power era. Though living in a northern European country not noted for sunlight, the Germans built several large power plants between 2002 and 2003.

In April 2003, the world's largest photovoltaic plant was connected to the public grid in Hemau near Bavaria. The peak power of the "Solarpark Hemau" plant is 4MW. With the support of Germany's EEG (a renewable energy law), this plant was followed by many larger systems in 2004. Germany's aggressive FiT program led the country into becoming the world leader of installed PV systems. According to the European Photovoltaic Industry Association, Germany installed 3.8gW by the end of June of 2010 and 2.13gW in June alone (PVResources 2011).

Asia's Technological Tigers—China, Japan, and South Korea—are rapidly overtaking Germany as the world's leader in PV technology and installed capacity. In Japan, a well-coordinated effort to fund research and development of PV technology has led to increased residential use of solar energy.

In the United States, the photovoltaic market amounted to 440MW peak in 2009. Of this, PV plants accounted for 372MW, with a peak output of more than 200kW each (PVResources 2011). Current trends indicate that a large number of PV power plants will be built, particularly in the south and southwest, where there is ample land in the sunny deserts of California, Nevada, and Arizona. Large properties are being bought with the aim of building more utility-scale PV power plants in these areas.

The latest innovation in putting energy generation onsite is taking place in Berkeley, California. The city has developed a public solar policy, called

Berkeley First, which allows building owners to avoid high upfront costs by amortizing the cost of systems over 20 years and paying for them through property taxes.

Another way to use the sun's energy is through solar thermal systems. These use sunlight to heat oil or water. Solar thermal systems are available on a small scale for individual consumers. There are many solar thermal systems installed as pool heaters or water heating systems for the residential market.

On a large scale, thermal technology is used for utility-size power generation. The large-scale systems use mirrors or lenses to focus solar energy on the liquid. The heated liquid then drives a turbine that generates electricity. These concentrated solar power (CSP) systems are of four basic types: trough, Liner Fresnel Reflector, tower, and dish. The most common is the trough system, which uses parabolic mirrors that concentrate solar heat on a fluid-filled receiver that runs the length of the trough (Nelson 2011).

The operation of a CSP solar plant requires large tracts of land and substantial volumes of water to provide cooling for the steam turbine. Many of the prime locations for these systems are remote, placing a burden on transmission lines and connection activities.

Although both PV and CSP are dependent on solar power input, CSP systems are easily fitted with thermal storage systems such as molten salt. Adding a storage component allows for operation at night or on cloudy days and turns solar energy into a more consistent power resource.

The United States pioneered CSP technologies, and there are a number of large solar thermal installations in California, Nevada, and now Hawaii. In mid-2010, the United States produced more than half of all solar thermal power in the world, although Spain is rapidly building solar energy plants (Nelson 2011).

Use of solar energy is increasing rapidly. In 2000 there were only 170MW of solar power generated globally. In 2010 the global market reached 20gW of installed solar power capacity, according to Greentech Media. They anticipate that the global PV market will reach 25gW by 2013, and come close to 100gW by 2020. They also anticipate that a number of large solar firms will approach the 1gW-capacity threshold in 2011, including SunPower, Yingii, Trina, SolarFun, and LDK. Most of these companies are Chinese; China now leads the market in solar cell and module vendors (Wesoff 2010).

While the United States doesn't have a coherent national energy policy, individual states are starting to enact their own laws. California leads the nation with a number of progressive policies. Colorado, Arizona, and New

York have policies aimed at increasing solar power generation. Also, it seems that utility-scale solar is taking hold. A report, prepared by research and publishing firm Clean Edge, found that solar power could grow to provide 10 percent of U.S. power by 2025. The report projected that nearly 2 percent of the nation's electricity would come from concentrating solar power systems, and 8 percent would come from solar photovoltaic systems.

The report noted that as solar power has been rapidly expanding, the cost per kilowatt hour of solar photovoltaic systems has been dropping. At the same time, electricity generated from fossil fuels is becoming more expensive. As a result, the report projects that solar power will reach cost parity with conventional power sources in many U.S. markets by 2015 ("Report: Clean Energy Trends 2011" 2011).

President Obama's Solar America Initiative, in collaboration with the U.S. Department of Energy, has targeted grid parity, where the electricity cost based on renewable energy production is the same as the cost of traditional coal-fired power. But to reach the 10 percent goal, solar photovoltaic companies must also streamline installations and make solar power a "plug-and-play" technology. It must be simple and straightforward to buy the components of the system, connect them together, and connect the system to the power grid.

Electric utilities need to take advantage of the benefits of solar power, incorporate them into future smart grid technologies, and create new business models for building solar power capacity. The report calls for establishing long-term extensions of today's investment and production tax credits, creating open standards for connecting solar power systems to the grid, and giving utilities the ability to include solar power in their rate base.

PV systems can be integrated into a building during construction. This fast-growing segment of the solar industry includes incorporating PV solar panels in building elements such as roofs, window overhangs, or walls. This reduces the material costs of the building construction and the installation cost of the PV panels. Passive solar building design can also take advantage of solar energy, using windows and interior surfaces to regulate indoor air temperature. As the price of solar panels and solar power generation drops, solar energy becomes a major component of the GIR.

New technologies and scientific advancements offer the keys to a new way of life. In this future scenario, most rooftops will have PV systems. The windows of office buildings will be covered with a thin film that, in reality, is another PV system. If the sun is shining in your neighborhood and your PV system is generating more electricity than you are using, you can store it in your car's battery, or pass it on to the neighbor a few blocks away who needs it. The GIR offers a whole new era of flexible energy generation and solar is a key element.

Geothermal

Geothermal energy creates power by extracting heat that is stored in the Earth. Heat was created and captured inside the Earth when the planet was formed. Heat is also created by the radioactive decay of minerals and solar energy absorbed at the surface. The Earth's heat has been used for space heating and bathing since ancient Roman times, but is now being used to generate electricity.

Most geothermal energy comes from heated water, or hydrothermal resources, that exist where magma comes close enough to the Earth's surface to transfer heat to groundwater reservoirs. This produces steam or high-pressure hot water. If the reservoir is close enough to the surface, a well can be drilled and the steam or hot water can be used to drive a turbine. The steam or hot water can also be used as a heat source. If the water temperature is moderate enough, it can be used directly to heat buildings, or for agriculture or industrial processes.

The first geothermal electricity generator was brought online in 1922 at The Geysers in Northern California. Other power plants were added and this geothermal field is now the largest in the world. In all, the United States has 77 geothermal power plants that provide about 4 percent of America's renewable electricity. The United States leads the world in geothermal electricity production, with 3,086MW of installed capacity (Geothermal Energy Association 2010).

In 2010 the International Geothermal Association (IGA) reported that 24 countries have a total of 10,715MW of geothermal power online, which is expected to generate 67,246gW of electricity. This represents a 20 percent increase in online capacity since 2005. IGA projects growth to 18,500MW by 2015, due to the projects presently under consideration (Geothermal Energy Association 2010).

Geothermal electric plants were traditionally built at the edges of tectonic plates, where high temperature geothermal resources are available near the surface. Recent technological advances have dramatically expanded the size of the resources that can be tapped and the range for the geothermal energy. This is especially important for applications such as home heating, and opens the potential for widespread exploitation.

Geothermal power is cost effective, reliable, sustainable, and environmentally friendly, and while geothermal wells release greenhouse gases trapped deep within the earth, these emissions are much lower per energy unit than those produced by fossil fuels. As a result, geothermal power has the potential to help mitigate global warming, if used in place of fossil fuels.

While the Earth's geothermal resources are huge, only a very small fraction may be profitably exploited. Drilling and exploration for deep resources is expensive, making this resource a limited one for the future. In the last few years, engineers have developed remarkable devices, such as geothermal heat pumps, ground source heat pumps, and geo-exchangers, which gather ground heat to provide heating for buildings in cold climates. Through a similar process, they can use ground sources for building cooling in hot climates. In the United States more communities with concentrations of buildings, like colleges, government centers, and shopping malls, are trying to adapt geothermal systems to maximize energy efficiency.

Biomass: Recycled and Reusable Generation

Biomass may be the oldest of all sources of renewable energy, having been around since the ancients learned the secrets of fire. Today, most people think of biomass as garbage because it can use dead trees, yard clippings, or even livestock manure to generate power.

Solar energy is stored in this organic matter, so under pressure and over time, a chemical process takes place that converts plant sugars into gases. The gas can then be burned as ethanol or used to generate electricity. The process is referred to as digestive, and it's not unlike an animal's digestive system. The appealing feature of this process for energy generation is that it can use abundant and seemingly unusable plant debris—rye grass, wood chips, weeds, grape sludge, and almond hulls.

To create usable energy from biomass, materials like waste wood, tree branches, and other scraps are burned to heat water in a boiler. The steam is used to turn turbines or run generators to produce electricity. Biomass can also be tapped right at the landfill or waste treatment plant, by burning waste products.

Biomass can produce energy without the need for burning. Most garbage is organic, so when it decomposes it gives off methane gas, which is similar to natural gas. Pipelines can be put into the landfills to collect the methane gas, which is then used in power plants to make electricity.

Animal feed lots can process manure in a similar way using anaerobic digesters. The digesters can create biogas from the manure, which is then burned to produce power. For example, dairy farms can use methane digesters to produce biogas from manure. In turn, the biogas can be burned to produce energy or used like propane. Biogas can also be derived from poultry litter.

Although biomass is a renewable energy source, the combustion process creates pollution. Biomass resources also vary by area and depend on the conversion efficiency to power or heat. For these reasons, many countries see biomass as a transitional renewable energy source, for use as they look for new technologies to either convert the emissions or process waste differently. For example, Denmark creates much of their energy with biomass but they plan to start converting and limiting its use totally in the next 5–10 years.

On the positive side, biomass raw materials get their energy from the sun and regrow quickly. Through photosynthesis, plants use chlorophyll to convert carbon dioxide into carbohydrates. When the carbohydrates are burned, they release the energy captured from the sun. Energy crops can be grown on marginal lands and pastures or planted as double crops. While most scientists say that making ethanol from corn is not efficient, it has been produced in the Midwest for years. Converting sugarcane to ethanol is considerably more efficient, and Brazil has adapted sugarcane to ethanol conversion as a major fuel source for transportation.

The Union of Concerned Scientists cites Minnesota's Koda Energy plant as an excellent example of generating energy from biomass. It is a combined heat and power (CHP) plant that uses biomass to generate renewable electricity as well as waste heat from the boiler. In 2009 Koda began generating electricity from oat hulls, wood chips, prairie grasses, and barley malt dust. About 170,000 tons of these agricultural wastes are used a year (Union of Concerned Scientists 2010).

Since the beginning of time, humans have captured the energy from biomass by burning it to make heat. In the First Industrial Revolution, biomass-fired heat produced steam power, and more recently this biomass-fired steam power has been used to generate electricity. Advances in recent years have shown that there are even more efficient and cleaner ways to use biomass.

Water as Energy: From Hydroelectric to Ocean Waves to Run of the Rivers

Throughout the world, water is moving. Gravity is always at work, moving water through tidal action, and pulling water from the mountains to the sea. For decades, humans have dammed rivers to generate electricity. Dams like the Grand Coulee Dam across the Columbia River in Washington, the Hoover Dam across the Colorado, and the Three Gorges Dam across the

Yangtze River in China, are some of the world's most imposing structural engineering accomplishments.

Hydroelectric power has been one of the key features of Western industrialization in the 2IR. However, there is a growing awareness that river damming destroys multiple ecosystems that cannot be recovered, and public resentment to dam construction is increasing. China, for example, has built many large dams to control water and rivers. One company now leads all of China in renewable energy system installations and is expanding globally. Yet, there are far better, cheaper, and more environmentally sensitive ways to generate electricity from renewable sources than building large dams.

There are four basic types of marine water currents: oceanic, tidal, wave action, and river. Water movement offers a huge potential for sustainable and consistent energy generation. Through tidal action, the world's oceans are constantly moving, providing the potential for enormous amounts of renewable energy. While there is a broad array of concepts and designs for capturing marine energy, there has not been the standardization needed to develop commercial viability. Eventually, as the technology advances and the world embraces more environmentally benign power generation, the needed market transformation will take place.

Tidal energy has been used by humans for centuries. There are European tide mills that date to A.D. 787. Medieval tide mills consisted of a storage pond, filled by the incoming tide through a sluice, and emptied during the outgoing tide through a water wheel. The tides turned waterwheels, producing mechanical power to mill grain.

Traditional tidal electricity generation involves the construction of a barrage, or barrier, across an estuary to block the incoming and outgoing tide. The dam includes a sluice that is opened to allow the tide to flow into the basin. The sluice is then closed, and as the sea level drops, the elevated water drives turbines to generate electricity.

Although the technology required to harness tidal energy is well established, tidal power is expensive, and there is only one major tidal energy generating station in operation. This is a 240MW system at the mouth of the La Rance river estuary on the northern coast of France. The La Rance generating station has been in operation since 1966 and has been a very reliable source of electricity. La Rance was supposed to be one of many tidal power plants in France, until their nuclear program was expanded in the late 1960s. Elsewhere, there is a 20MW experimental facility at Annapolis Royal in Nova Scotia, and a 0.4MW tidal power plant near Murmansk in Russia.

There are several other potential tidal power sites worldwide, including constructing a barrage across the Severn River in western England. Similarly, several sites in the Bay of Fundy, Cook Inlet in Alaska, and the White Sea in Russia have been found to have the potential to generate large amounts of electricity. The waters off the Pacific Northwest are ideal for tapping into an ocean of power using newly developed undersea turbines. The tides along the Northwest coast fluctuate dramatically, as much as 12 feet a day, and have exceptional energy-producing potential. On the Atlantic seaboard, Maine is also an excellent candidate for tide-generated power.

Tidal energy could be an environmentally friendly, renewable source of electricity. High construction costs are the main barrier to increased use of the tidal energy. Capital costs for tidal energy projects are also high, and construction can take as long as 10 years.

The technology required for tidal power is well developed. Offshore tidal power generators use reliable hydroelectric generating equipment, conventional marine construction techniques, and standard power transmission methods. An offshore system eliminates several environmental and economic problems that have prevented the development of commercial-scale onshore tidal power plants. Three offshore tidal projects are in development in Wales, where tidal ranges are high, renewable power is a strong public policy priority, and the electricity marketplace gives it a competitive edge.

The demand for electricity on an electrical grid varies with the time of day. The supply of electricity from a tidal power plant will never match the demand. But, due to the lunar cycle and gravity, tidal currents—although variable—are reliable and predictable and their power can make a valuable contribution to an electrical system. Tidal electricity, when used instead of fossil fuel to generate electricity, can reduce emissions of greenhouse and acid gasses.

In several parts of the world, small power turbines are mounted along rivers. As the water flows, the turbines generate electricity. This is being done in the EU, where the water generates a considerable amount of energy without harming the surrounding land or changing the natural elements in the water.

This could easily be done in the United States. A large amount of U.S. electricity is consumed in a state that borders an ocean or a Great Lake. This means that electricity based on these offshore hydrokinetic resources could easily provide substantial amounts of power to high-demand electricity markets.

For the Planet's Benefit and Human Health: Renewables Have to Come First

The shift from a fossil fuel, nuclear power–based economy to a renewable energy–centered economy has begun. China, Japan, South Korea, and the European Union are well into the process. China, Japan, and South Korea (unlike the United States) have national energy policies, programs, and financing. The EU has taken a regional approach and is pushing its members toward energy independence through renewable energy generation and the use of storage devices and other technologies.

World nations are coming to realize that there is a clear need for a consistent, intelligent, and long-term energy policy that will stabilize the energy markets. There is also a whole new era of job creation and business opportunities available to nations with the political foresight to take advantage of GIR opportunities. So far, America clearly lags in this effort while China has embraced it. Other countries, like Chile, have taken notice and are working quickly to power their energy-intensive economies with renewable energy.

The new global economy is being led by what scholars characterize as social capitalism. This is apparent in China, which has an enormous opportunity to push the GIR forward. The new economy of the GIR will be a green one that is focused on sustainable development for the public good. It will focus on renewable energy generation and encompass the social infrastructures of energy, transportation, telecommunications, buildings, and natural resources.

Renewable energy is available in multiple ways—wind, solar, geothermal, marine, and biomass are the most common and most available. Wind, sun, and tides are ancient processes that have been used to generate energy for most of human history. They are readily available, highly efficient, and environmentally friendly. Their use produces little or no increase in GHGs, global warming, or acid rain. Most importantly, they are sustainable and not subject to dwindling supplies, extraction problems, price disruptions, or political leverage. Their energy can be produced on utility scale, community scale, village scale, individual facility scale, or even a nano scale.

Renewable Energy Systems Are Protecting the Environment and Changing Local Communities

Converting to renewable energy sources will require a significant financial and political commitment. The United Nations estimates that in 2009 there was $162 billion invested in renewable energy worldwide. About

$44 billion was spent in China, India, and Brazil, collectively, and $7.5 billion in the many poorer countries (UN Environment Programme 2009).

Although the United Nations and other world bodies have urged developing nations to look first at environmentally friendly renewable energy instead of carbon- and fossil-based fuels as they add capacity, it is extremely difficult to build any system that is not grid-connected. Relatively speaking, it is easier to build $300 million solar systems in California than wind farms or industrial-size solar systems in developing countries, even when they feed into a grid for distribution.

Painstakingly slowly, the GIR is seeping into the nongrid world of the rural poor. Small self-contained solar and biomass power generation systems are being used to create enough power for a house or a hamlet. As small-scale renewable power becomes cheaper and more dependable, it is providing the first access to electricity for people far removed from the power grids that serve the cities of developing countries.

In these rural villages, two modern megatrends—globalization and the GIR—are coming together to change lives. In Africa, small Chinese-made solar power systems are appearing, balanced precariously atop of tin-roofed shacks far away from power grids. In other areas, simple subterranean biogas chambers make fuel and electricity from cow manure. In Nepal, mini-hydroelectric dams derive enough energy from a local river to power a village. In India, Husk Power Systems has installed 60 systems that use rice husks to generate power for 250 hamlets (Nerenberg 2011).

Solar systems allow people in remote areas to use cell phones. In Kenyan villages, owning a cell phone is a lifeline. By selling a goat or two, a villager can buy a tiny solar system for $80. Coupled with a $20 cell phone, a villager can make small money transactions, check the price of chickens, or connect with family members who have moved to the city. In areas without electricity, charging a cell phone can require a long walk or bicycle ride to a shop where the phone can be charged. Some shops are so backlogged that it takes days to charge a phone. Under these conditions, the cell phone is a precarious lifeline at best. Now, with the small solar powered systems, the cell phone can be charged at home, and one villager can develop a small business charging the cell phones of others.

The United Nations estimates that 1.5 billion of the world's 6.8 billion people live without electricity. As many as 70–80 percent of the population in nations like Kenya (where people still cook and heat with primitive fuels like wood, cow dung, or charcoal) are without electricity (International Energy Agency 2010). The explosion of cell phone use in these rural villages is the major driver for these small cheap solar systems. In an area without banks, the cell phone is embraced as a tool for commercial transactions and personal communications.

In poor, rural areas, cheap solar systems are playing an epic, transformative role. These tiny power centers are allowing the rural poor to save money on candles, batteries, kerosene, and wood, which in many parts of the world is becoming less available as governments try to halt the deforestation of rural areas. Babies are no longer inhaling the fumes of smoky kerosene lamps. Solar panels coupled to a car battery and a powerful, low wattage LED lamp can provide lighting. After sundown, older children can study their numbers and do their reading lessons just like their cousins in the city. People who live in mud-walled huts and raise goats to survive are using cell phones for financial transactions and as a link to the modern world.

Solar systems are being combined with low-wattage, high output LEDs. The Firefly LED system consists of a solar panel that can be placed in a window, and is connected to a LED desk lamp and phone charger. Larger Firefly systems can run radios or small appliances.

In some African countries, villagers use underground tanks in which manure from cows is converted to biogas. Then the gas is pumped through a rubber tube to a burner for cooking. In other areas, the biogas is used to generate electricity.

These and other tiny, individual, renewable power-generating systems are slowly becoming available in undeveloped areas with the help of nonprofit agencies and the United Nations. However, they buck the traditional grid-connected power mold, and investors are reluctant to put money into fragmented markets consisting of poor rural consumers.

The benefits of the GIR are coming slowly, but they are coming to the poor. In Africa, there is evidence that a true market is emerging for home-scale renewable energy and low-consuming appliances. As costs keep coming down, families are willing to sell a goat or borrow money from a relative in the city to buy a tiny solar system. The cell phone is a must-have for African villagers, and this passion is driving the need for tiny, nongrid, renewable-energy systems. It is this nexus of digital electronics and renewable energy that is at the forefront of the GIR.

Developing an economy based on renewable energy sources is precisely what is needed to create the jobs that the American economy so desperately needs to wrench itself out of the worst recession since the Great Depression. A new renewable power industry, with all the ancillary activities like smart grids, energy efficient buildings, and new generations of efficient equipment and automobiles, will revitalize U.S. innovation, capital investment, and business practices. The retraining of the workforce for green jobs will give new life and purpose to a moribund higher education system that is self-indulgent and lacks rigor compared to the Asian and Indian systems.

As the GIR becomes robust, and more and more nations embrace it, renewable energy generation from sustainable sources will become the norm. A few decades from now, people will look back with contempt at the way the 2IR lifestyle squandered fossil fuels, destroyed ecosystems with coal extraction, and threatened the health and survival of the planet through ruthless disregard of global warming and climate change.

Chapter 8

Smart Green Grids in the GIR

Every day, Americans flip a switch to light their kitchen and make their morning coffee. It is a habit that requires no planning or forethought, yet it triggers the flow of electricity that overlays modernity and defines conventional everyday life. To reach the kitchen, electricity must first flow across hundreds of thousands of miles of power lines and pass along a series of stations and substations. This electrical transmission network is called a power distribution grid and it is a technological marvel of the 2IR. However, a few modifications could make it the high-tech brain of the GIR.

Most people are familiar with the conventional central power plants of the 2IR as described in chapter 1. They burn coal, oil, or natural gas, or are powered by nuclear or hydroelectric power and are housed in large concrete structures with massive silos and pipes usually surrounded by high wire fences and security guards, located long distances from population centers. Several classic American power plants are visible from the New Jersey turnpike. Hundreds of thousands of motorists pass them every day without a thought as traffic pours through the Holland Tunnel into lower Manhattan. Similar large power plants are located across the United States. Many were constructed decades ago in once isolated areas, but are now surrounded by housing developments and massive building complexes. Few large hydroelectric dams have been built in the United States since the Great Depression, due to concerns about their environmental impact.

All 2IR central power plants share a similar concept: fuel in the form of coal, natural gas, oil, hydroelectric power, or a nuclear reaction is used to create heat that drives a steam turbine, which generates electrical power.

Dr. Jerry Jin helped with the data and insights on smart green grids (see chapter 12 in Clark 2010).

Coal, oil, and natural gas are burned to heat a steam turbine that powers a spinning electrical generator. Dams usually spin a water wheel that, in turn, spins the electrical generator. A nuclear reactor creates heat for the steam turbine, which drives the electrical generator.

Electricity is produced in different forms. Large electrical generators naturally generate what is called three-phase alternating current (AC) power, as opposed to direct current (DC) power. Most household electrical appliance use single-phase 120-volt AC power. Batteries with an electrical-chemical reaction generate DC power. Large-scale power plants produce three different phases of AC power at the same time; each phase is offset 120 degrees from the other. There are four huge wires exiting from every power plant, one for each phase and one for the electrical ground that is common to all three.

Transmitting electricity requires additional steps. Three-phase power leaves the generator and enters a transmission substation, where large transformers boost the electricity's voltage for long-distance travel on the transmission grid. There is a transmission line for each phase of the three-phase AC electricity, plus ground lines. The transmission lines are usually held aloft by huge steel structures and a typical transmission distance is 300 miles.

Electricity must then be distributed to the end users. The electricity produced by these 2IR systems is so powerful that it must be "stepped-down" along the distribution grid to eventually provide the 120 AC power connected to that switch used to light the kitchen. The conversion from power "transmission" to power "distribution" takes place through a power substation. The substation uses transformers to break the electricity down, and at the same time uses a "bus" to split the power off in multiple directions. From the distribution bus, it goes along distribution wires (three—one for each phase, plus lines for grounds). It will then go through regulator banks and finally a tap.

The final step brings electricity into individual homes. Since a house only needs one of the three-phased wires, the tap separates the distribution wires and brings one into a transformer drum. A transformer drum breaks the voltage (usually 7,200) down to the 240, which is common to most homes and offices. The 240-volt line is then attached to a watt-hour meter, which measures the amount of electricity used. Then it goes to a circuit box that separates the electricity into 120-volt lines, and one or two 240-volt wires for heavy-duty machines like washers and dryers.

Thomas Edison started America's electricity distribution grid in New York City in 1882. His entrepreneurial goal was to create a company that met the individual demand for electricity with power from one central power plant. Edison built the Pearl Street Station, which provided electricity to 85 customers. This new "grid" technology was immediately popular. Demand grew and investors raced into the market. It was a wild time with high demand driving fierce competition, and by 1922 there were 3,774 different companies producing and distributing electricity all over the United States. The Great Depression forced consolidation of these independent companies, and eventually 85 percent of the U.S. electric industry was controlled by16 companies. The U.S. power grid today works in the United States with oversight from the Federal Energy Regulatory Commission (FERC). In September 2011, FERC's chairperson remarked on how antiquated the energy system in America is, after a massive power outage in Southern California, Arizona and norther Mexico.

Most of the U.S. grid was built during this pre–World War II period in classic radial patterns. The individual grids look much like a tree, with large transmission lines forming the trunk, connecting to larger branches (substations), and moving along smaller and smaller lateral branches as they extend outward. To provide more reliability, the grids include mesh networks that provide redundancy. Redundancy allows power to be rerouted during a failure, while the damaged lines are being repaired.

Today's U.S. power grid is a massive relic with overlapping networks of lines. More than 500 companies own parts of it, with FERC providing regulatory oversight by regions. Overall, it covers about 180,000 miles, which is equivalent to traveling back and forth from San Francisco to New York 60 times (Brian 2000).

The size, age, and design of the grid make it vulnerable to problems. A large failure in one area can cascade along the grid, causing widespread power outages, as occurred in August 2003 when 50 million Americans and several million Canadians lost power during the worst blackout in United States and Canadian history. The cascading power outage hit Toronto first, then struck Rochester, Boston, and finally New York. It took just 13 minutes for the blackout to spread throughout the 80,000 square-mile Canada–United States Eastern Interconnection power grid ("Major Power Outage Hits New York" 2003; Clark and Bradshaw 2004).

The cause of that massive blackout was not known until years later, after a thorough investigation. It seems that a tree fell on a substation, which triggered a failure that then ricocheted around the northeastern United States and southern Canada (Clark and Bradshaw 2004). The resultant disaster and loss of lives motivated energy experts to rethink the central power grid for both reliable energy supply and national security.

America is not the only place where widespread blackouts have occurred. During that same summer of 2003, France experienced a similar blackout in which several thousand people died due to the lack of power for health equipment and cooling systems. Northern Italy also had a massive energy blackout that summer, although theirs was more inconvenient than disastrous.

The process triggered by that simple act of flipping on the kitchen light is as critical to understanding the 2IR as it is to contemplating the GIR. In fact, that example underlines an issue at the core of both industrial eras: whether electricity is generated using the "dirty" fossil fuels of the 2IR or the "green" renewable fuels of the GIR, it must flow efficiently from the transmission site to the end user. In the 2IR, power distribution grids are central to a nation's infrastructure and key to commerce, communication, and almost every facet of an electricity-driven lifestyle. In the GIR, the generation of energy from local onsite renewable power sources creates a sustainable system that is essential to future survival.

Regulation and reliability became greater concerns after the August 2003 northeast United States and southern Canada energy meltdown. Massive central power grids are regulated and supervised by federal (FERC), state (independent system operators), and local governments (municipal utilities). The U.S. power grid is over a hundred years old. It was the result of the 2IR and is a tribute to 2IR technology. Across this enormous, old, cumbersome, and technologically outdated distribution system, electricity generation and consumption must be balanced because energy—unlike water—is not stored, but is consumed almost immediately after being produced. However, the United States still has no national energy plan, only massive central plants and their associated power transmission lines.

To maximize the GIR, the United States must replace fossil fuel generation with renewable energy generation. This will help mitigate global warming and preserve the environment. Hence, the United States needs to move to a local onsite power grid and distribution network that is technologically "smart." The need for smart green grids, to generate large amounts of power and meet the needs at the local onsite level, is fundamental to developing the modern energy networks in every grid-connected nation.

Efficient Power Transmission and Distribution

The term "smart grid" (sometimes known as "intelligent" or "wise" grid) started to appear at the end of the 20th century during the dot-com era. Basically, it refers to using digital or information technology to control and enhance the electricity networks that form a power grid. Originally, these systems used land lines, such as telephone wires, but today the systems may be wireless or WiFi-based.

Early proponents of smart grids thought that by using two-way digital communications to monitor energy use, large amounts of electricity could be saved and costs could be reduced. They thought that smart technology would also increase reliability and transparency. The idea was to overlay the current electrical grid with an information system that included smart meters at the consumer's end. The utilities believed that smart meters, with real-time use capability, would help motivate consumers to use less power and be more efficient with electricity. Smart meters would also provide the utilities with a more accurate, real-time, and immediate customer cost profile.

In 2003 the Pacific Northwest National Laboratory produced a groundbreaking study for the U.S. Department of Energy called *GridWise®: The Benefits of a Transformed Energy System*. The report described how electricity would continue its historical growth trend far into the future. The report's authors concluded that this growth would require added generation, transmission, and distribution, costing up to $1.4 billion/gW ($1,400/kW in year 2000 dollars) on the utility side of the meter. "The amount of capacity needed in each of these categories must supply peak demand and provide a reserve margin to protect against outages and other contingencies. Thus, traditional approaches to maintaining the adequacy of the nation's power generation and delivery system are characterized by lower than desirable asset utilization, particularly for assets located near the end-user," said the report (Pacific Northwest National Laboratory 2003).

The report turned out to be timely. In a remarkable bit of prescience (the report was published before the summer of 2003's electricity meltdown in the northeast United States), it mentioned that, "The disparity between current levels of investment in generation and transmission suggests a looming crisis that creates a strong element of urgency for finding alternative solutions" (Pacific Northwest National Laboratory 2003). California had gone through a similar energy crisis with its rolling brownouts during 2000–2002. The authors noted that the current power supply infrastructure had been created to serve load as a passive element of the system. However, they emphasized that information technology was at the point of being able to create an integrated system that could include superior control and protection functions as well as enable real-time economic interaction with the grid.

According to the report, the GridWise concept would offer extraordinary benefits. There would be improvements in the generation, transmission, and distribution components of the power grid, as well as in the customer sector. In a stunning conclusion, the report noted that the:

> total potential benefit of implementing these technologies over the next 20 years is conservatively estimated to have a present value (PV) of about $75 billion. When estimated on the basis of a less conservative implementation scenario, the PV of these benefits is shown to essentially double . . . While implementation costs were not considered and the error band on the total benefit value is likely to be large (possibly ± $25 billion), the major conclusion of this exercise is that the GridWise concept has the potential for great economic value and should make a major contribution to transforming the present electric generation and delivery infrastructure into the power grid of the future. (Pacific Northwest National Laboratory 2003)

Understandably, the report triggered lots of interest, and the utilities began pushing for smart end-user meters that could provide accurate, real-time customer costs. At the same time, the lure of a huge nationwide market caught the attention of the world's large IT and computer firms, which were interested in developing grid software and pushing to make the old grid smart. Major American high tech companies (Cisco, Microsoft, Google, Yahoo, HP, Apple, etc.) got involved in both research and implementation of the new smart green grid. However, it was primarily in the EU and Asia that nations and communities began installing such systems.

Today, the term "smart green grid" covers both GIR modernization of the transmission of power and local onsite distribution grids, while "smart grid" usually refers only to modifications of the massive central power grids from the 2IR. The hope is that smart green grids would increase competition between providers, enable increased use of variable energy sources, provide automation and monitoring capabilities, and enable the use of market forces to drive energy conservation.

The first step in developing a smart green grid is to create a smart meter. Smart meters are high-tech electricity meters that use digital communication so that the utilities can interact with the consumer. With smart meters, utilities can let customers know when they will be charged variable electric rates that reflect the cost differences of generating electricity during peak or off-peak periods. Smart meters would also let utilities interact directly with consumers' appliances. For example, the utility would be able to control a homeowner's water heater and have it turn on when electricity was cheaper to produce.

Globally, there is growing support for smart meters. Many nations, including the United States, are now subsidizing the installation of the devices for central power plants as the first step in creating a smart grid. European Union politicians are pushing to connect 80 percent of the region's homes and businesses to smart green meters by 2020 (Scott 2009). Emerging giants like India and China plan to install the technology in new buildings, thus combining both central power plants and onsite energy systems.

It may have been simple theft more than anything else that pushed the adoption of smart meter technologies. According to U.S. industry analysts, rampant power theft and fraud was the driving force behind Italy's widespread adoption of smart meters (Scott 2009). In less than a decade, Italy has become the world leader in the development of a smarter electrical grid (the Italians also saw the need for, and are very involved in, the GIR). Regional governments set up programs, which provided funding for smart green grids. For example, Northern Italy reacted to the 2003 summer power blackout in a proactive manner to get control over their power demands as well as to become far more active in addressing local environmental protection and climate change mitigation needs.

According to *Bloomberg Businessweek*, about 85 percent of Italian homes are now fitted with smart meters—the highest percentage in the world, with more such devices than exist in the entire United States (Scott 2009). In 2001 Italy's dominant national utility, ENEL, started a five-year program to install smart meters for its customer base of 40 million homes and businesses. ENEL cited a variety of reasons for the new program, including improving efficiency and helping customers reduce energy bills.

ENEL invested $3 billion in the project, which included meters that send usage readings automatically to the central office and display time-of-day pricing to customers. The Italian utility can now collect customer data and manage its energy network remotely, instead of sending out costly technicians. The improved data on consumer electricity use helps ENEL run its power plants more efficiently. The utility says that the annual cost savings is about $750 million, which will allow it to recoup the infrastructure investment in just four years (Scott 2009).

According to ENEL, their customers now have greater control over their energy bills. The meter is typically installed in a convenient place in the home like a kitchen cupboard or laundry room. When electricity prices are high, for instance during the peak evening period or on cold winter nights, the smart meter informs household members of higher rates. This allows them to alter their habits, such as postponing a load of laundry until the next morning, to avoid higher charges. Analysts figure that attentive ENEL customers have been able to cut their bills by as much as one-half by closely watching energy prices and usage (Scott 2009).

The worldwide installed base of smart meters is expected to triple to 180 million units by 2014, according to ABI Research. The EU (with its population nearing 500 million) has set mandatory installation targets for 2020. The EU is aiming to have smart meters installed for 64 percent of the customer base (a goal of 115 million smart meters). North America may have 45 million units installed by 2014 (Scott 2009).

Smart meters are just part of the story. An electrical grid is a collection of multiple networks and multiple power generation companies. Within the grid, various operators are employing different levels of communication and coordination, most of which are manually controlled. Smart grids increase the connectivity, automation, and coordination between these suppliers, consumers, and networks that perform either long distance transmission or local distribution tasks. A smart grid national system would have thousands of digital transformers and grid sensors in substations and buildings, making them smart green grids. Like the Internet, it would use routers and pathways to control the traffic. The key benefit to smart grids is their capacity to monitor and control energy consumption in real time. Smart grids are also much more secure and far less vulnerable to external interference.

To maximize the green benefits of the GIR, the smart green grid needs to seamlessly integrate renewable energy. Solar and wind power are sustainable, abundant, affordable, and make good use of natural resources. These green technologies have to replace fossil fuel power and become the drivers for the new power distribution grid. However, the lack of a new grid system that can effectively mix large amounts of renewable energy—solar and wind—into the system creates a bottleneck to large-scale renewable generation. That holds back the rapid development and full potential of renewable energy.

In a fully functioning GIR smart grid, the grid will also be green, which means that the use and monitoring of power will not come from the central grid, but from local power generation. Currently, when solar panels are installed on a home or building, the power is green and used locally. The system owner receives a credit for at least a portion of the power they generate themselves. This process is called "net metering."

With net metering, the system owner does not use the power directly. For example, when a building owner installs solar panels on a building's roof the panels are connected to the grid through the meter. When the panels generate power, the power is not routed directly to the building's lights or appliances, but onto the grid. As the power goes back to the grid, the utility company credits the building owner with the kWh being generated.

The system owner does not generally get credit for all the power generated. In California, for example, the owner only gets credit for 90 percent of the building's previous year consumption. So owners erect systems that are sized to generate no more than 90 percent of the kWh the building used last year, regardless of the availability of roof space, or the willingness of the owner to purchase a larger system.

The utilities make the case that the old grid system is unable to adequately function as a two-way transmission system. It was built to move electricity in a linear fashion from power plant to building, not from building to power plant. Proponents of smart grids say that is precisely what is needed. They say that a truly smart grid would be able to monitor all the electricity on the system, all the time. If the solar panel system on a building with an exceptional southern exposure on a perfect sunny day is generating more kWh than the facility is using, then the smart grid should be able to slide the excess electricity to a nearby building with an exposure that is less advantageous. Then, on days that are cloudy, the system would provide energy from an alternative source, from a storage device, or from several storage devices.

This scenario is technologically possible, but there are numerous obstacles. Throughout the United States, there are millions of square feet of available rooftop space that could accommodate solar panels. With panel pricing continuing to drop, and unlimited quantities of energy available from the sun, it would be relatively easy to generate more than enough energy for the nation. However, this would mean that there would be millions of small power-generating systems that would feed into the grid, each system controlled independently. This type of system bypasses the large investor-owned utilities (IOUs), which now own the large power generation plants. Currently, at least in California, where the state is pushing for more renewable energy, the IOUs are proposing to build massive solar farms in the desert, and then ship the electricity hundreds of miles to the end user. This idea has created enormous opposition from the state's environmental groups, and seems contrary to the goals of renewable energy and to the technological capabilities of a smart grid.

Electricity must be moved from where it is generated to where it is used by consumers of all kinds and sizes. Each step along the way lowers the power efficiency, degrades the lines, and creates more operation and maintenance costs. Generating power onsite is more efficient and much cheaper.

One of the key differences between 2IR fossil-fuel power and GIR renewable energy power is that energy generated from renewable sources is less robust than that generated from fossil fuels. It doesn't have the same "oomph" coming out of the power plant, and therefore doesn't travel as well. For this reason, it is important to place renewable energy generation systems close to urban end users to maximize the usable power. This is why building solar farms, even concentrated solar installations, hundreds of miles from end users is not as efficient as smaller roof systems that feed power directly to the user.

Solar and wind renewable energy technologies are developing quickly and are rapidly becoming economically competitive. Wind systems are getting lighter, less expensive, and more durable. Their cost per kWh is on par with many conventional fossil fuel–driven systems, especially if environmental cleanup costs are included. The cost of solar panels is plunging and solar electricity systems will most likely be commercially successful in five years, even without government rebates. However, the lack of smart grid infrastructure forms a bottleneck to the investment in new renewable generation. It also limits the ability to maximize the full power of renewable energy (such as that which could be provided by utility-scale solar electricity systems).

A new grid system would enhance demand response and energy efficiency, and would be compatible with large supplies of renewable energy. The new systems would place digital transformers and grid sensors in homes and create utility substations to enable a grid-smart data system. The move to advanced energy technology requires a partnership between energy efficiency and renewable energy, controlled and distributed by a smart green grid.

Data Response and Power Transmission Lines

Power system management and optimization is really about data management, response, and efficiency. Smart systems assure that when the need for electricity reaches a significantly high level or energy reduction is needed, customers are notified, which alerts them to begin energy conservation behaviors. This reduces demand and enhances the system's reliability. This type of system is consistent with our nation's security concerns and supports the renewable power standard.

One of the major obstacles to a national smart grid is the problem of replacing many miles of aging transmission lines. The current U.S. power infrastructure uses a grid protocol from century-old technology. There are limitations for interstate transmission lines that have resulted in stalling thousands of MW of wind projects and solar power deployments. In the United States, for example, long distance power transmission has been the major barrier to the success of renewable power standard implementation in certain regions. Challenges in financing, permitting, and pricing transmission systems have created nearly insurmountable obstacles in many cases.

The energy industry envisions a transformation similar to that created by the Internet. The smart grid would be an active system like a nerve network that determines, responds, and controls the power needed for consumers. The network control system would operate on a national or regional scale to dispatch energy, manage the energy flow protocol, and maintain distribution control. The network would be self-healing and able to respond to problems quickly and efficiently. For example, a smart network control system would respond to a power block by circumventing it to initiate recovery.

The Internet uses the concept of distributed control for information exchange. The Internet has a central hub or plant, where a web-host computer or a designated computer server acts autonomously under a global protocol. A smart grid can help identify power losses as well as greater efficiencies. It can manage its effective response to consumers. Internet web technology could be used to manage the power grid, which would benefit consumers by reducing costs.

For energy consumers and power generation owners, buyers, and sellers, using the nerve network of the Internet with a smart grid would increase flexibility for the power purchase transaction. Consumers could choose more efficient energy options over more costly ones.

The existing power grid has specifications that match the current coal-based power plants. Additional transmission lines will be required for electricity transmission within each interconnected region and among the three power interconnections. Getting renewable energy sources connected to the grid and adding additional transmission lines as required will produce sufficient power flow to consumers. A new power grid can adapt to cleaner and more efficient power plants.

The need for a smart grid is fundamental to developing modern energy networks in every grid-connected nation. Governments are making large-scale investments, and IOUs, like Italy's ENEL, are making huge commitments. Tech giants like Cisco are planning to bring products to the smart grid market. In a May 2009 announcement about its smart grid roadmap, Cisco identified the expanding smart grid market as one of its new market priorities, valued at $100 billion dollars. The advancement of these products will come to address the challenges of market demand, clean energy need, and greenhouse gas emission reduction.

The smart green grid is ideally suited to meet the challenging demands in the production, distribution, and use of electricity. The challenge is to take a century-old power-grid infrastructure, turn it upside down, connect it to numerous renewable energy sources, and manage it digitally. Smart green grids hold the key to adopting and using renewable energy power generation systems.

Transformation of the mainstream energy market requires an advanced grid infrastructure with superior energy efficiency and green technology. Thankfully, the adoption of smart green grids has begun, with European leadership. It is spreading throughout the world's power distribution industry and is becoming a fast-growing technology sector. While it is going to take a sustained commitment by regulators, state lawmakers, utilities, and other stakeholders to change an old relic to something that is efficient, flexible, and technologically "smart," the United States must do it if the nation wants to keep up with Europe and Asia and reap the benefits of the Green Industrial Revolution.

Chapter 9

Emerging Commercial Technologies Empower the GIR

Just as the First and Second Industrial Revolutions gave birth to extraordinary new ideas, the Green Industrial Revolution is spawning truly remarkable technologies. Many of these wonders are being developed in the United States, which has always been a leader in technological innovation. However, when new GIR technologies emerge in the research laboratories of American universities and companies, their commercialization is often stalled or stunted by the entrenched interests of the 2IR. It is a two-step process: first, the invention, and second, market transformation.

American dependency on and government support of fossil fuels has slowed the nation's adoption of renewable energy, fuel cells, and other GIR technologies. These technologies include "green" (environmentally sound) fuels that contain no fossil fuels or derivatives for use in vehicles and buildings. Related technologies, such as regenerative braking (which uses the friction caused by braking to create electricity) and the storage devices needed to store power, have not yet made it into commercial low-cost use (at least not as long as costs are measured using traditional business economic analyses). In many instances, the technological skills needed to develop environmentally sound technologies are not supported or endorsed; emphasis is instead placed on traditional 2IR skills, technological training, economics, and business development.

As in most innovative eras, when startling new technologies are created, some businesses and members of the public reject them as too strange or too costly. Apple Computer proved that this is not always the case. Under the leadership of Steve Jobs, Apple consistently managed to commercialize new technologies, despite high introductory costs. An exploration of a few GIR technology wonders provides a sense of what inventive engineers and scientists can do when faced with truly monumental societal, financial, and 2IR challenges.

The Green Industrial Revolution needs to start with energy conservation and efficiency. This focus has started in the 2IR, but has fallen far short of achieving movement into the GIR.

Energy Conservation and Efficiency Technologies

By far the biggest challenge in moving from the 2IR to the GIR concerns how energy is produced and then distributed. By the end of the 20th century, energy was being produced and distributed through central power plants. Because fossil fuels were a core component to the 2IR, the central plants were primarily powered by coal and then oil and natural gas combustion. Around the plants, the environment was often polluted. Getting coal or other fossil fuels to the plants required long-distance transportation by trucks, trains, and, in many cases, ships. Liquefied natural gas (LNG) is an example of a fossil fuel from the 2IR that is shipped across vast stretches of oceans from one country to another, then offloaded, stored, and piped long distances across land. Transportation of fossil fuels not only adds costs, but also creates dangers for people living near the transport routes, adds potential environmental threats, and even presents national security risks.

It was in the interests of the companies that mined coal or drilled oil and gas to supply fossil fuels for power plants and the vehicles that transported the fuels. These companies used their enormous political power to obtain subsidies and tax breaks to build and operate their plants. For example, ExxonMobil reported over $45 billion in profits for 2010, yet paid no U.S. federal tax. Furthermore, company executives grew rich while the workers near the plants were paid low wages and often became ill from air and water pollution. However, these power plants were located far away from urban areas and more often in poor economic areas. Over time, the polluted land around these plants became cheaper and was sold to low-income families.

Then in the United States, the GIR started to emerge. The first renewable energy systems were primarily wind farms and solar concentrated systems located long distances from the buyers and end users. Long transmission lines continued to be a major problem, due to inefficiencies, environmental concerns, and the high cost of transporting energy over power lines.

Today, these long-distance renewable systems are slowly changing to onsite energy generation. Through government programs, some power utilities and states offer incentives to consumers who install onsite solar and wind energy systems. Usually, these tax breaks must be renewed annually or semiannually, which limits their impact. Nonetheless, more and more solar panels are being installed on rooftops, and more wind farms are being built in large agricultural areas of the U.S. Midwest. Small wind turbines are even becoming part of building designs. This onsite power generation creates the need for new technologies to store the energy for when the sun has gone down or the wind is not blowing.

Government-led initiatives are encouraging the EU and Asia to research and develop storage devices from fuel cells to flywheels, and to create systems that better integrate renewable energy power sources. Government initiatives also put more pressure on consumers to conserve energy and become more efficient. This helps drive the development of new technologies for lights, smart meters, and grids to maximize energy use and efficiency. The United States is starting with programs such as Energy Star (the U.S. Department of Energy ranking system) and the LEED standards from the U.S. Green Building Council, to promote better energy management as a first step toward the GIR.

The U.S. Department of Energy estimates the nation uses about 20–30 percent of its total electrical consumption to light facilities (Brodrick 2006). This includes exterior parking and street lamps. For most of the 20th century, lighting was from incandescent bulbs and later fluorescent tubes.

Incandescent bulbs make light by heating a metal filament wire to a high temperature until it glows. As a legacy technology, incandescent light bulbs use a large amount of electricity and are being replaced by newer technologies that improve the ratio of visible light to heat generation. The EU is phasing out incandescent light bulbs in favor of more energy-efficient lighting such as LED and low–energy use systems. The United States plans to phase out incandescent light bulbs by 2014, and Brazil has already phased them out. California is leading the way in the United States by prohibiting the sale of incandescent bulbs and fluorescent tubes after January 2010 (California State Law, Assembly Bill No. 1109 2007).

After World War II, fluorescent tubes gradually replaced incandescent bulbs in commercial office facilities and high-use residential areas. A fluorescent lamp or fluorescent tube is a gas-discharge lamp that uses electricity to excite mercury vapor. A fluorescent lamp is more efficient than an incandescent lamp in converting electrical power into useful light. A fluorescent lamp fixture requires a ballast to regulate the current through the lamp. Fluorescent lights were first available to the public at the 1939 New York World's Fair. Improvements since then include better phosphors, longer life, more consistent internal discharge, and easier-to-use shapes. Eventually, more efficient electric ballasts started to replace the older magnetic ballasts, which eliminated the odd humming noise.

The lighting industry is quiet and low-key, so new technological advances are not announced with great fanfare. However, the GIR's digital communications and Internet connectivity have created a whole new generation of commercial lights that are energy efficient, dimmable, and most importantly, able to adjust to exterior daylight or high peak load demands. These transformative technologies are being created by electrical engineers intent on providing the perfect lighting at all times. The perfect energy-efficient commercial lighting is a combination of building design and task-ambient lighting placement. The essence of this new generation of lighting is the Internet-connected dimmable ballast, working in combination with a high-output, low-wattage fluorescent tube.

In the energy and utility industries, the goal is to provide just enough electricity to meet the end user's demand. In the 2IR system, the central power grid delivers a constant stream of electricity, whether the end-user needs that much power or not. Green Industrial Revolution efficiencies come from building end-user equipment that can respond to the ebb and flow of consumer and grid demand, rather than providing a constant stream of electricity that may not be needed during non–intensive use periods. For example, why use a constant pulse of electricity to light an office that is flooded with daylight in the morning hours, but shadowed in the afternoon? Ideally, the sun would be used for lighting in the morning, and electricity would be used in the afternoon. The new generation of lighting will respond to these changes by using sensors on the window glass. The sensors will determine the light level and transmit the information through an Internet connection to a sensitive dimmable ballast that will then provide just enough electricity to power the lights at the optimal level for worker comfort and productivity.

The move toward increased efficiency is not so much about workplace comfort as it is about utilities clinging to the remnants of their 2IR central power authority. The 2IR central grid systems are nearing capacity in many major metropolitan areas in the developed world. Manhattan's grid is at capacity. The grid in California's Silicon Valley is near it. Additional electricity is needed to power new businesses and an increasing number of electronic devices (phones, computers, televisions, etc.), but the grids and generation systems are at their maximum capacities. The only way to get quick relief is through conservation and an energy efficiency program that reduces demand at a pace that is faster than the growth of new commerce.

Utilities have difficulty responding to large swings in demand. For example, HVAC systems demand more power on hot summer afternoons. Demand also rises in the evening, when people come back from work. In high-humidity places like Dallas, Texas, demand is fairly constant during a workday, when many people are in offices. Under these conditions, the grid delivers a steady stream of electricity. Then, around 5 P.M., workers begin to flee the office and head home (and in Texas, the homes are large). People walking into a large house at the hottest time of the day immediately turn their air conditioning units to full blast throughout the house. Several million units surging at the same time sends the grid and power supply into overload. These demand extremes can cause power interruptions or failures. Utility companies go into a panic at this point because power interruptions or failures are expensive, time consuming, and focus consumer attention on them.

The easiest way to prevent surges, or overload, is to keep everything in balance. So, when demand spikes, electricity is delivered, but not at such a high level. There comes a point that, to avoid major problems, electricity delivery has to be rationed, which means that some people have to get less. In this world, rationing doesn't work, so the utilities, both in America and other developed nations, have been pushing what is called demand response, or "peak-load management," which offers the ability to reduce electricity use during extreme demand times.

Peak-load management tries to transfer some of the electricity use from high-demand periods to low-demand periods. Originally, peak-load strategies centered on offering utility rate discounts to customers willing to reduce their energy use during surge time periods, referred to as "incidents." For example, a utility program might offer a large industrial customer a 5 percent rate discount if it reduces electricity use during days that are declared incident days. Incident days may happen 10–15 times a year in some areas. Once notified of an incident day, the customer would turn off lights or machinery, or reduce the amount of air conditioning, to meet the agreed-on reduction in electricity use. If the customer fails to meet the reduction, then a penalty is accessed.

Notification of an incident day was initially done by phone, then through an email to the customer. Now, with the emerging technologies of the GIR, peak load management is rapidly becoming more flexible and more effective. Utilities are offering deeper discounts to lure customers into letting the utilities connect directly to the customer's energy management system (EMS). The EMS is connected via the Internet to the facility's dimmable ballasts for the lights, and the large central HVAC systems for air conditioning. In this way, the utility can automatically reduce a customer's systems to meet the agreed-on level.

Eventually, as the new technologies come into the marketplace, simple equipment like washing machines and dryers will have Internet connectivity and smart operating systems. They can then be controlled remotely, and the utilities will notify the machines not to operate at peak load periods. While most people would consider it intrusive, this sort of mass-marketplace energy management will go a long way toward optimizing and balancing energy use throughout a smart grid system that integrates renewable energy generation with other technologies.

In the 1960s, scientists developed a revolutionary, low-wattage light bulb called LEDS for light-emitting diodes. LEDs. Early LEDs were practical electronic components that emitted low-intensity red light. Today's LEDs can provide light across the visible, ultraviolet, and infrared wavelengths, with very high brightness.

LEDs are an extraordinary new generation of lighting. A 6-watt LED can provide the same amount of light as a standard 60-watt commercial overhead interior light. LEDs have a longer lifetime (measured in years, instead of months), improved robustness, smaller size, faster switching, and greater durability and reliability. As a new technology, they are more expensive than traditional light bulbs, but the price is dropping quickly as new manufacturers come to market.

The long life of an LED light is a huge advantage to a commercial building operator, yet creates a dilemma for the LED manufacturer. With prices headed below $20 a bulb, manufacturers are in panic because each LED bulb may last 20 years, greatly reducing the need to buy new ones. Companies like industry leader Lemnis Lighting are considering creating purchase agreements for consumers willing to pay for lighting as a service, rather than having to replace light bulbs continually. These companies would include maintenance and upgrades as an incentive.

Technological advances in lighting may seem mundane compared to some of the other GIR emerging technologies, but lighting impacts every office, home, and room in the modern world. It can even be life changing. In a mud-walled shack in a rural village in an undeveloped part of the world, lighting can offer a child a chance to study his or her lessons and slip the bonds of sustenance farming.

Hydrogen and Fuel Cell Storage

A fuel cell is a storage device that has an electrochemical cell that converts a source fuel into an electrical current. It generates electricity inside the cell through reactions between a fuel and an oxidant, triggered in the presence of an electrolyte. Unlike a conventional battery, fuel cells consume reactant from an external source, which must be replaced but lasts considerably longer. They are also more likely to be environmentally sound, in terms of their manufacturing and disposal. The reactants flow into the cell, and the reaction products are separated and flow out of it, while the electrolyte remains within it. The electrolyte provides the electric power. Fuel cells can operate continuously, as long as the reactant and oxidant flows are maintained.

In 1839 British scientist Sir William Grove invented the first fuel cell. Grove knew that water could be split into hydrogen and oxygen by sending an electric current through it, a process called electrolysis. He hypothesized that by reversing the procedure, you could produce electricity and water. He created a primitive fuel cell and called it a gas voltaic battery. Fifty years later, scientists Ludwig Mond and Charles Langer coined the term *fuel cell* (National Institute of Standards and Technologies 2005).

A fuel cell produces an electrical current that can be directed outside the cell to do work, such as powering an electric motor or illuminating a light bulb. Fuel cells can power an engine or an entire city. Because of the way electricity behaves, the current returns to the fuel cell, completing an electrical circuit. The chemical reactions that produce this current are the key to how a fuel cell works. There are several kinds of fuel cells, and each operates a bit differently. But in general terms, hydrogen atoms enter a fuel cell at the anode, where a chemical reaction strips the atoms of their electrons. The hydrogen atoms are now "ionized," and carry a positive electrical charge. The negatively charged electrons provide the current through wires to do work.

Oxygen enters the fuel cell at the cathode, where it combines with electrons returning from the electrical circuit and hydrogen ions that have traveled through the electrolyte from the anode. Whether they combine at anode or cathode, hydrogen and oxygen combine to form water, which drains from the cell. As long as a fuel cell is supplied with hydrogen and oxygen, it will generate electricity. Even better, since fuel cells create electricity chemically rather than by combustion, they are not subject to the thermodynamic laws that limit a conventional power plant. This makes them more efficient in extracting energy from a fuel. Waste heat from some cells can also be harnessed, boosting system efficiency still further.

The basic workings of a fuel cell are not that complicated, but building inexpensive, efficient, reliable fuel cells has proved difficult. Scientists and inventors have designed many different types and sizes of fuel cells in the search for greater efficiency. The choices available to fuel cell developers are constrained by the choice of electrolyte. For example, the design of electrodes and the materials used to make them depend on the electrolyte. Today, the main electrolyte types are alkali, molten carbonate, phosphoric acid, proton exchange membrane (PEM), and solid oxide. The first three are liquid electrolytes; the last two are solids.

The type of fuel also depends on the electrolyte. Some cells need pure hydrogen, and therefore demand extra equipment such as a "reformer" to purify the fuel. Other cells can tolerate some impurities, but might need higher temperatures to run efficiently. Liquid electrolytes circulate in some cells, which requires a pump. The type of electrolyte also dictates a cell's operating temperature. For example, "molten" carbonate cells run hot, just as the name implies. Each type of fuel cell has advantages and drawbacks when compared with the others, and none is yet cheap and efficient enough to widely replace traditional ways of generating power, such coal-fired, hydroelectric, or even nuclear power plants.

Some fuel cells depend upon fossil fuels, greatly reducing their value as a GIR technology. In 2010 the television news show *60 Minutes* featured the Bloom Energy fuel cell, which uses natural gas as a fuel source. While the Bloom fuel cell has low-carbon emissions and high efficiencies, it still uses natural gas—a fossil fuel with emissions and particulates. Bloom makes a 100kW solid oxide fuel cell that sells for about $700,000. After incentives, Bloom claims its server generates power for 9–11 cents per kWh, a calculation that includes fuel, maintenance, and hardware expenses. Bloom is renowned for its customer base of high-tech and environmentally sensitive corporate clients like Wal-Mart, Google, and FedEx (Boudway 2011). Yet, because Bloom's fuel cell uses natural gas to generate electricity, it is a fossil fuel–based system from the 2IR.

The real goal in the fuel cell industry is to perfect a hydrogen-based fuel cell that electrolyzes renewable energy to take the place of the gasoline-based internal combustion engine. Hydrogen fuel cells use hydrogen as the fuel and oxygen as the oxidant. For more than three decades, the Los Alamos National Laboratory (LANL) and other U.S. Department of Energy (DOE) national research labs have been investigating how to use hydrogen fuel cells for transportation, industry, and homes (see the Hydrogen Association of America, http://www.hydrogencarsnow.com/).

Most hydrogen cars use fuel cells to generate electricity and electric motors to power the car. A few use internal combustion engines modified to accept hydrogen and burn it as fuel. And some use a hydrogen compound to generate hydrogen-on-demand to power the vehicle. Much research, development, and money is being poured into hydrogen fuel cell research, as this is seen as the ultimate in green-car technology.

Every car company has developed a hydrogen fuel cell vehicle and they are all planning to market them, starting in the EU and Japan. In the United States, at least nine major car manufacturers are introducing such cars in 2015, under lease agreements rather than for sale. This pre-GIR strategy is partly to control the market, but also to monitor and measure performance, and focus attention on the need for refueling stations. Within the next decade, these cars will be sold and there will be refueling stations in the homes of the hydrogen fuel car owners (Hydrogen Association of America, http://www.hydrogencarsnow.com/).

Hydrogen offers the promise of a zero-emission engine, where the only byproduct created is a small amount of environmentally friendly water vapor. Current 2IR fossil fuel–burning vehicles emit pollutants such as carbon dioxide, carbon monoxide, nitrous oxide, ozone, and microscopic particulate matter. Hybrids and other green cars address these issues to a large extent, but hydrogen cars are the only ones that don't produce any pollutants. The U.S. Environmental Protection Agency estimates that fossil-fuel automobiles emit 1.5 billion tons of greenhouse gases into the atmosphere each year (Hydrogen Association of America, http://www.hydrogencarsnow.com/). Switching to hydrogen-fuel-based transportation would eliminate this serious cause of climate change.

Roughly speaking, the energy output from one kilogram of hydrogen is equivalent to the energy output of one gallon of gasoline. However, the production cost of the one kg of hydrogen is considerably lower. Typically, a gasoline internal combustion engine (ICE) with a mechanical drive train is 15–20 percent efficient, while a hydrogen ICE is about 25 percent efficient. Hydrogen fuel cell vehicles with electric hybrid drive trains can be up to 55 percent efficient, or about three times better than today's gasoline-fueled engines. Because the production of hydrogen (by steam reformation of natural gas or electrolysis of water) is expected to be about 75–85 percent efficient, the net energy efficiency of hydrogen fuel cell vehicles will still be better than twice that of gasoline ICE vehicles. The delivered price of hydrogen is approximately $3 per kilogram.

In 2005 Honda leased the first commercial hydrogen car (to a family in Redondo Beach, California). In 2008 the company created the first production-line hydrogen fuel cell car (built in Japan): the Honda FCX Clarity. It is powered by a 100kW V-Flow fuel cell stack, a lithium-ion battery pack (50 percent smaller than the one on the previous FCX), a 95kW electric motor, and 5,000 psi (pounds per square inch) compressed hydrogen gas storage tank that yields a range of 270 miles. In 2011 the Honda FCX Clarity was available for lease only to customers in Southern California. Customers need to live near one of the active hydrogen fueling stations in Torrance, Santa Monica, Culver City, or Irvine. Nevertheless, the GIR may finally be starting in the United States, at least in Southern California. Almost all cars makers have hydrogen fuel cell cars for the mass market and plan to lease then sell them in 2015. The process has already begun in some select cities around the world.

There are dozens of other GIR hydrogen auto prototypes being road-tested across the world. In fact, Sweden and Norway are working to make these Scandinavian countries the first in Europe to construct a working, public, hydrogen highway system, complete with fuel cell cars and hydrogen refueling stations. A public-private partnership called Hydrogen Sweden is promoting hydrogen as a green energy carrier for cars and is working to develop a public refueling infrastructure. Hydrogen Sweden, founded in 2007, is a nonprofit organization that currently has 40 members including Honda, BMW, Volvo, StatOilHydro, H2 Solution, Air Liquide and Arise Windpower (Hydrogen Association of America, http://www.hydrogen carsnow.com).

Hydrogen should be produced from renewable resources. One method is to use biomass, although the process emits some carbon dioxide. Hydrogen can also be derived by using wind, hydroelectric, or solar power to electrolyze water. Today, electrolysis is still expensive, but companies in Canada and Norway predict rapidly declining costs. Other companies around the world are developing hydrogen systems for stationary power (fixed energy generation facilities such as power plants). Further reductions in production costs through lower cost electrolyzers and the use of low-cost, off-peak, renewable electricity, could dramatically reduce the future cost of electrolytic hydrogen.

Elemental hydrogen (as opposed to liquid hydrogen) is not a viable energy resource. Hydrogen from solar, biological, or electrical sources requires more energy to make than is obtained by burning it. However, hydrogen can be used as an energy carrier, like a battery.

Elemental hydrogen has been widely discussed as a possible carrier of energy on an economic scale. Used in transportation, hydrogen would burn relatively cleanly, with some NO_x emissions, but without carbon emissions. The infrastructure costs associated with full conversion to a hydrogen economy would be substantial. However, if refueling was done in the home or workplace using water or other renewable sources, electrolyzers could produce the hydrogen needed for fuel cells.

Today, industrial production of hydrogen is mainly from the steam reforming of natural gas and, less often, from more energy-intensive hydrogen production methods, such as the electrolysis of water. However, hydrogen from electrolysis and other renewable energy sources is slowly gaining momentum. Many GIR countries around the world are making plans to provide it commercially for vehicles in 2015 inline with the global marketing timetable plan developed by the major auto makers.

Fuel cell technologies, particularly hydrogen fuel cells, are an integral part of the GIR; they offer an exceptionally attractive alternative to oil dependency and fossil fuels. Scientists and manufacturers have a lot of work to do before fuel cells become a practical alternative to current energy production methods, but with worldwide support and cooperation, a viable hydrogen fuel cell–based energy system may be a reality in a few years.

Regeneration Braking: From Trains to Cars to Trains and Back Again

When a vehicle brakes to slow down, the reduction in speed creates kinetic energy. With conventional braking systems, excess kinetic energy is converted to heat by friction in the brake linings. As a result, the energy is wasted. Regenerative braking systems transform kinetic energy into another form of energy, which can be saved in a storage battery. This energy recovery mechanism is used on hybrid gas and electric automobiles to recoup most of the energy lost during braking. The stored energy is then used to power the motor whenever the car is in the electric mode. The most common form of regenerative braking is used in hybrid cars like the Toyota Prius, and involves using an electric motor as an electric generator.

Regenerative braking is emerging as a viable technology for electric railways. For railways, the generated electricity is fed back into the onboard energy supply system, rather than stored in a battery or bank of capacitors, as is done with hybrid electric vehicles. Energy may also be stored using pneumatics, hydraulics, or the kinetic energy of a rotating flywheel.

Regenerative braking systems are not new. They were used by Louis Antoine Krieger in the late 19th and early 20th centuries as front-wheel drive conversions for horse-drawn cabs. The Krieger electric landaulet had a drive motor in each front wheel, with a second set of parallel windings for regenerative braking. The Raworth system of regenerative control was introduced in England in the early 1900s. It offered tramway operators economic and operational benefits.

Tramcar motors worked as generators and brakes, by slowing down the speed of the cars and keeping them under control on descending gradients. The tramcars also had wheel brakes and track slipper brakes, which could stop the tram, should the electric braking systems fail. Following a serious accident, an embargo was placed on this form of traction in 1911. Twenty years later, the regenerative braking system was reintroduced, but failed to become the dominant technology in the 2IR because of the political and economic influences of the oil, gas, and internal combustion vehicle industries.

Despite resistance, regenerative braking has been in limited use on railways for many decades in the 2IR. For example, the Baku-Tbilisi-Batumi railway (Transcaucasian railway or Georgian railway) started using regenerative braking in the early 1930s. This technology was especially effective on the steep and dangerous Surami Pass. In Scandinavia, the Kiruna to Narvik railway carries thousands of tons of iron ore from the mines in Kiruna in the north of Sweden down to the port of Narvik in Norway. These trains generate large amounts of electricity with their regenerative braking systems. For example, on the route from Riksgränsen on the national border of Sweden to the Port of Narvik, the trains use only a fifth of the power they regenerate. The regenerated energy is sufficient to power the empty trains back up to the national border. Any excess energy from the railway braking is pumped into the power grid and supplied to homes and businesses in the region, making the railway a net generator of electricity (Jno, Robertson, and Markham 2007).

The first U.S. use of regenerative braking was the 1967 AMC Amitron. The American Motor Car company developed an energy regeneration brake for this concept car. The AMC Amitron was a completely battery-powered urban car, with batteries that were recharged by regenerative braking, which increased the range of the automobile. After the Japanese commercialized the technology, Ford and the Chevrolet licensed it back from Toyota for their hybrid cars. Regenerative braking is also used in the Vectrix electric maxi-scooter.

High-Speed Rail and Maglev Trains Have Become Realities

Magnetic levitation, or maglev, may do for the GIR in the 21st century what airplanes did for the 2IR in the 20th century. Most Asian and EU countries have developed and implemented high-speed rail train systems. Not the United States. Now the GIR countries are developing maglev or floating high-speed train systems. Maglev train systems use powerful electromagnets to float the trains over a guideway, instead of the old steel wheel and track system. A system called electromagnetic suspension suspends, guides, and propels the trains. The system uses magnetic levitation (guided and controlled tension from a large number of magnets) for lift and propulsion along a track (National Labor Report 2010).

Maglev trains do not need an engine and, therefore, produce no emissions. They are faster, quieter, and smoother than conventional systems. The power needed for levitation is usually not a large percentage of the overall consumption. In fact, most of the power is used to overcome air drag, which is a factor with any high-speed train (National Labor Report 2010).

Maglev technology is based on a 1934 patent, and was pioneered by German Transrapid International after World War II. Transrapid completed the first commercial implementation for the Chinese in 2004 with the Shanghai Maglev Train, which connects the city subway network to the Pudong International Airport. This system transports people more than 19 miles in just over seven minutes. In 2010 the Chinese started a new project to extend the line to Hangzhou, about 105 miles away, and construction should be completed in 2014. The proposed speed is over 200 miles per hour, which would allow the train to travel the distance in 27 minutes. The line will become the first inter-city maglev rail line in commercial service in the world and also the fastest inter-city train. The Chinese have plans for similar maglev trains throughout the country. They reason that going from one city to another via a maglev train is far easier, more efficient, uses less fuel, and is better for the environment than any other form of transportation except the bicycle.

Flywheels and Storage Devices

Universal's Incredible Hulk roller coaster ride uses so much electricity that if it did not have multiple large flywheels to store energy, the company would have to invest in a new substation and risk brownouts on the local grid every time the ride launched. Voted the Number 1 roller coaster by Discovery Channel viewers in 1999, the Incredible Hulk roller coaster features a rapidly accelerating, uphill, full-speed launch, as opposed to the typical gravity drop. Powerful traction motors throw the car up the track, and to achieve the brief, but very high, current required to accelerate a roller coaster train to full speed uphill, the park uses several motor generator sets attached to large flywheels to store the energy.

Flywheels are an old technology that has roots in the Neolithic spindle and the ancient Greek potter's wheel. The potter's wheel features a heavy round stone, connected to a pedal that is pumped by the potter. The flywheel stores the fluctuating pedal movements as inertia, and creates a smooth, steady, turn of the wheel. Used in the Middle Ages in water wheels, flywheels were also used by James Watt and James Pickard in steam engine applications to help launch the First Industrial Revolution. Flywheels resist changes in their rotational speed, which steadies the rotation of the shaft.

Through the wonders of mechanical engineering, this old technology is being refitted as a modern-day energy-storing device. Flywheel energy storage (FES) works by accelerating a rotor, or flywheel, to a very high speed and maintaining the energy in the system as rotational energy. When energy is extracted from the system, the flywheel's rotational speed is reduced; adding energy to the system correspondingly increases the speed of the flywheel.

Most FES systems use electricity to accelerate and decelerate the flywheel, but devices that use mechanical energy are in development. Advanced FES systems have rotors made of high-strength carbon filaments. The filaments are suspended by magnetic bearings and spin at speeds from 20,000 to over 50,000 rotations per minute (rpm) in a vacuum. Such flywheels can come up to speed in a matter of minutes—much quicker than some other forms of energy storage.

Over the past two decades (the end of the 2IR), scientists studied flywheels extensively and created ways to use them as power storage devices in vehicles and power plants. Flywheels can be used to produce high-power pulses in situations where drawing the power from the public network would produce unacceptable spikes. A small motor can accelerate the flywheel between pulses. Another advantage of flywheels is that it is possible to know the exact amount of energy stored by simply measuring the rotation speed.

Flywheel technology can be used as a replacement for conventional chemical batteries. They have a long life cycle and require little maintenance. Flywheels are also less damaging to the environment, being made mainly from inert materials.

Flywheels can even power transportation systems. In Switzerland in the 1940s, OC Oerlikon created flywheel-powered buses called gyrobuses. In England, Parry People Movers, Ltd., created a railcar powered by a flywheel for the Stourbridge Town Branch Line. It went operational in 2010 with two units ("Parry People Movers" 2010).

FES can even help regulate the line voltage for electrified railways. This will improve the acceleration of unmodified electric trains, and increase the amount of energy recovered back to the line during regenerative braking, helping to keep costs down. Several large cities including London, New

York, Lyon, and Tokyo have pilot FES projects ("New York Orders Flywheel Energy Storage" 2009).

Current FES systems have storage capacities comparable to batteries and faster discharge rates. Many are used to provide load leveling for large battery systems, such as uninterruptible power supply systems for data centers, where they save considerable amount of space when compared to battery systems. On average, flywheel maintenance runs about half the cost of traditional battery systems. All that is needed is a basic annual preventive routine that includes replacing the bearings every five to 10 years (which takes about four hours). Newer flywheel systems completely levitate the spinning mass using maintenance-free magnetic bearings, thus eliminating mechanical bearing maintenance and preventing failures.

Biofuel: A 2IR Transitional Energy Power

Although renewable energy is the goal, transitional energy sources are an interim step. Biofuels (fuels made from living or recently living organisms) are examples because they can be burned and used in ways similar to fossil fuels, but they are not carbon-based. Ethanol from corn or sugar cane is an example of a biofuel. Unfortunately, it takes about the same amount of fossil-fuel energy to make corn-based ethanol, so there is no real benefit to its use as an alternative to gasoline. Sugar cane is more efficient, and it is widely used in Brazil.

The two most promising new sources for biofuels are algae and a process called metabolic engineering. Though both must be burned to create energy, they are significantly cleaner than fossil fuels as substitutes for gasoline and diesel, and can be sustainably produced. In one of history's most delightful ironies, metabolic engineering produces a clean fuel from switchgrass. This is the plant that the great herds of prairie bison fed on for centuries before America's Great Plains became Nebraska, Iowa, and Kansas, and were crisscrossed by the highways and corn fields used to support the fossil-fuel industry.

Algae as a Biofuel Source

Algae are a group of simple organisms that are among the world's most ancient creatures with a fossil record that goes back three billion years to the Precambrian era. The U.S. Algal Collection lists almost 300,000 specimens, ranging from one-cell organisms to large plants like the giant ocean kelp that grows to 150 feet in length. The glory of algae is that they are photosynthetic (able to use sunlight to convert carbon dioxide and produce oxygen), and "simple," because their tissues are not organized into the many distinct organs found in land plants (Walton 2008).

The U.S. National Renewable Energy Laboratory, with funding from the DOE, has been studying single-cell algae as a biofuel source since 2005 (Walton 2008). What has scientists and researchers excited about algae is their rapid growth cycle—up to 30 times faster than corn—and the ease with which they can be turned into lipids (a green, goopy vegetable oil). This oil, much like any vegetable oil, can be burned and used as a substitute for carbon-based diesel oil or corn-based ethanol. In a strict sense, burning algae or other biofuels does not reduce atmospheric carbon dioxide, because any CO_2 taken out of the atmosphere by the algae is returned when the biofuels are burned. However, it does reduce the introduction of new CO_2 by reducing the use of fossil hydrocarbon fuels.

Algal oils have many attractive features. They can be farmed easily through algaculture, using land that is not suitable for agriculture. They do not affect fresh water resources and can be produced using ocean and wastewater. They are biodegradable and relatively harmless to the environment if spilled.

At current production costs, oil made from algae is more expensive than other biofuel crops such as corn, but could theoretically yield between 10 and 100 times more energy per unit area. One biofuel company claims that algae can produce more oil in an area the size of a two-car garage than soybeans can produce in an area the size of a football field, because almost the entire algal organism can use sunlight to produce oil (Walton 2008). The DOE estimates that if algae fuel replaced all the petroleum fuel in the United States, it would require just 15,000 square miles of farming area, which is only 0.42 percent of the nation's land mass. This is less than one-seventh the area that is currently used to grow corn in the United States. The U.S. Algal Biomass Organization claims that algae fuel can reach price parity with oil in 2018, if granted production tax credits (Feldman 2010).

While much of the research on algae is focused on creating oils, either for food or as a transitional fuel for vehicles, one Canadian cement company has discovered a unique application. Ontario's St. Mary's Cement plant is using algae from the nearby Thames River (running through Ontario) to absorb carbon dioxide. The plant started a pilot project in 2010, using algae's photosynthesis to absorb the carbon dioxide produced during cement manufacturing. Martin Vroegh, the plant's environment manager, says the algae project is believed to be the first in the world to demonstrate the capture of CO_2 from a cement plant (Hamilton 2010).

Through this process, the St. Mary's plant is turning CO_2 into a commodity rather than treating it as a liability. The CO_2-consuming algae will be continually harvested, dried using waste heat from the plant, and then burned as a fuel inside the plant's cement kilns. Additionally, the goopy oil can be used as a biofuel for the company's truck fleet.

The company is preparing for a carbon-constrained future that will not treat cement makers and other energy-intensive industries kindly. "The amount of exposure to carbon pricing we face as an industry is very high," says Vroegh. "If we want to be around tomorrow we have to be sustainable. This project helps us achieve that" (Hamilton 2010).

Algae can be used to make vegetable oil, biodiesel, ethanol, biogasoline, biomethanol, biobutanol, and other biofuels. Algae-based oil holds great promise for a variety of products, ranging from jet fuel to skin care and food supplements. The potential for large-scale production of biofuel made from algae holds great promise, because algae can produce more biomass per unit area in a year than any other raw material (While 2011). The break-even point for algae-based biofuels should be within reach in about 10–15 years.

GIR Fuel from Plants

Some of the most advanced scientific minds in biology, chemistry, and now metabolic engineering are working to develop useful microbes that will break down simple plants into starches and sugars, and eventually into clean fuel. Grappling with this challenge are several American national laboratories and universities, including the Lawrence Berkeley, Lawrence Livermore, and Sandia national laboratories, and the University of California campuses at Berkeley, Davis, and San Diego (U.S. Department of Energy, Energy Information Administration 2011).

For over a century, scientists have made fuels and chemicals from the fatty acids in plant and animal oils. The hope is that a synthetic microbe can cost-effectively break down tough plant materials like wood chips and plant stalks and extract the simple sugars so they can easily be converted to fuel. The University of Illinois, Urbana, and the University of California, Berkeley, have a 10-year, $500 million grant program to develop algae and other biofuels.

While scientists are engineering fuel-producing microbes, farmers and agriculture experts in the Midwest are developing the inexpensive plants needed to produce biofuels. The DOE has tapped Tennessee's Oak Ridge National Laboratory (ORNL) and the Biofuels Feedstock

Development Program to lay the groundwork for this new source of renewable energy.

Perennial grasses are well suited for this purpose, particularly *Panicum virgatum* (switchgrass), which at one time stretched across America's Great Plains. It was a plant of the American prairie, growing throughout tens of millions of acres for centuries (before the coming of the white man, the railroad, and the steel plow). Switchgrass is not suburban lawn grass. It is big and tough, and it can stand 10 feet high, with stems as thick and strong as hardwood pencils. It grows fast, and is highly efficient, capturing solar energy and turning it into lots of cellulose that can be liquefied, gasified, or burned directly. It also reaches deep into the soil for water, and uses the water very efficiently. Because it spent millions of years evolving to thrive in the climate and growing conditions spanning much of America, switchgrass and other grasses like *Miscanthus* are remarkably adaptable.

Both perennial grasses are ideal for energy crops, and both are considerably better at producing energy than corn. Scientists estimate that if you include the energy required to make tractors, transport farm equipment, plant, and harvest, the net energy output of switchgrass is about 20 times better than corn (Samson and Stamler 2009).

Researchers are working to boost switchgrass hardiness and yields. At the same time, they are adapting varieties to a wide range of growing conditions, and reducing the need for nitrogen and other chemical fertilizers. By fingerprinting the DNA and physiological characteristics of numerous varieties, the researchers are steadily identifying and breeding varieties of switchgrass that show great promise for the future (Samson and Stamler 2009).

Most of America's former tall grass prairie has been planted in corn and beans, with switchgrass growing only in parks and preserves. Now, in research plots and laboratories in the Plains states and even in the Deep South, this is changing. The tall, native switchgrass of the prairie once fed millions of bison and was vital to America's ecological past. Now it may become vital to its economic future. Grown as an energy crop, switchgrass may help fuel millions of cars and trucks, spin power turbines, and supply chemicals to industries.

Commercializing Emerging Technologies

Truly innovative technologies are not transformative solely by themselves. Today's mass market is far too vast and complex to support these remarkable GIR technologies by traditional means. To be commercialized, these technologies need governmental support through a consistent and long-term incentive process from the research and development stage to public financing. For example, Thomas Edison was only able to establish a commercial electricity company when the costs could be supported by local governments, and electricity could be made available to the mass market at reasonable prices. Since the end of the World War II, the industrialized nations have all used government research and development monies to support the commercialization of everything from diesel fuels to the Internet. In fact, American government officials traditionally justify the funding of national labs, NASA, and even the U.S. Department of Defense, based on dual use or transfer of technologies.

Government incentives, tax breaks, and even procurement are critical to the commercialization of new technologies. Governments can also assist in the introduction of new technologies through regulations and standards. Today, the advancements of technology to speed communications and to slow climate change are all linked to government regulations and oversight. California has often led the way in this regard with its emissions controls, environmental laws, and atmospheric regulations.

Public and private partnerships can work collaboratively to create new industries and jobs. California used such a partnership for the zero-emission vehicle regulations introduced in the early 1990s with a focus on electric battery-powered vehicles.

Transforming markets is not easy, and it requires complete vertical integration. For example, creating broad market acceptance for dimmable ballast lighting technology will require training of hundreds of thousands of sales people and installers, and providing incentives to entice the customers. Yet, the results of a market transformation from current lighting technology to GIR new generation technology—LEDs, dimmable ballasts, demand response—could cut the U.S. electricity bill by 20 percent.

The United States did an extraordinary job in optimizing the fossil-fuel 2IR technologies with government support through public policy, research and development, subsidies, and supportive tax benefits. This was the powerful model that drove the U.S.'s prosperity during the 20th century. It is precisely the model being used by Europe and Asia to commercialize the technologies of the GIR.

Unfortunately for the United States, the nation's lifestyle, its government-funding processes, its politics, and its business and corporate practices are entrenched in fossil-fuel dependence.

The result is that American companies' ability to develop the emerging technologies of the GIR and then bring them to market has been restricted and handicapped. While the United States has invented some of the most popular early GIR technologies (like the regenerative braking system), they were sold to other companies and nations who foresaw the GIR and brought them to market. For over a decade, the U.S. auto industry could easily have manufactured electric and hydrogen fueled cars, but resisted production because of high profits from the sale of fossil fuel–based vehicles. Sadly, that shortsightedness led to the collapse of iconic U.S. automakers that had been the core of the nation's manufacturing sector for almost a century.

Chapter 10

The Next Economics

It's not all about money—or is it?

In light of the October 2008 world financial meltdown, it seems silly to think that the supply-side, deregulated, free-market economics so passionately espoused by President Ronald Reagan and Prime Minister Margaret Thatcher in the 1980s would work for a 21st century world threatened by irreversible environmental degradation. Even the bastion of supply-side economics, *The Economist*, ran a special issue in July 2009, almost a year after the global economic collapse, that discussed how modern economic theory had failed. The cover story showed the Bible melting and featured the headline: "Collapse of Modern Economic Theory" (2009).

The October 2008 economic implosion from trillions of dollars in credit swaps, hedge funds, subprime mortgages, and related marginal derivatives (which nearly pushed the Western world's financial structure into the abyss), underlined what happens when governments ignore their responsibility to govern. In the end, the worst financial disaster since the Great Depression was a testament to the venal side of free market capitalism—greed, stupidity, carelessness, and total disregard for risk management. Market economists and others had argued that there was no need for regulation. Government would act as "the invisible hand." These are not behaviors that can be repeated if the planet is going to survive climate change and its impact on the earth and its inhabitants.

Like the First and Second Industrial Revolutions, the Green Industrial Revolution must develop an economy that fits its social and political structures. The First Industrial Revolution replaced an agrarian, draft animal–powered economy with one powered by steam engines and combustion machine–driven manufacturing, an evolution that was accelerated by colonial expansion. The Second Industrial Revolution created a fossil fuel–powered economy that extracted natural resources in an unregulated, unfettered, consumer-fed, free-market capitalist society. The glories of the 2IR—market economics and liberal democracy—reached epic heights in the early 1990s with the end of the Cold War and the subsequent collapse of the Soviet Union and communist bloc.

Combined with the ideological triumph of democracy over communism were the extraordinary breakthroughs in information and digital technology driven by American innovation, science, and investment capital. California's Silicon Valley was, and to some extent continues to be, the center of the capitalist market-driven universe that was propelled by cheap fossil fuel, ingenuity, and intelligent and productive venture capital. American culture and the U.S. economic model spread as the world's regions and nations became more interconnected.

It was a glorious period for America, and it continued unabated until September 11, 2001, when 19 al-Qaeda terrorists hijacked four commercial passenger jets, crashing two into New York City's World Trade Center, and one into the Pentagon in Arlington, Virginia. Passengers in the fourth managed to overrun the terrorists, but then the plane crashed in a field near Shanksville, Pennsylvania.

Since then, free market economics and liberal democracy have been shaken by a series of events originating in the Middle East: al-Qaeda and Taliban insurgences in Afghanistan, the U.S. invasion of Iraq, terrorist attacks on civilians in Yemen, and Islamic terrorist bombings in England and Spain. As a result, the oil-rich Middle East has gone from the Western world's gas station to its battleground. The Middle East has become a volatile, transitional region driven by demands for democracy, modernity, and economic participation, contrasted with the fundamentalists' desire for self-identity and community. Many developed nations are embracing GIR economics just to reduce the leverage on Western politics and economics now enjoyed by the region.

The economics of the 2IR resulted in a widely disproportionate amount of money in the hands of the industries involved with fossil fuels and related products, as well as the manufacturers who were able to prosper by supplying cheap energy. It was wealth built at a significant cost to the health of the general population, which had to live with polluted air and water, climate change, acid rain, and waste accumulation. The public also has been forced to pay excessive taxes to subsidize the oil industry, which continues to be the wealthiest industry on earth. Michael Hoexter, in his "Cheap Energy Contract: A Critical Roadblock to Effective Energy Policy in the U.S.," argued that energy is kept artificially cheap to satisfy corporations and politicians (Hoexter 2008).

The real cost of energy has never been reflected in the market due to government financial support and promotion of oil, gas, coal, and nuclear power. For example, *Forbes* magazine reported in April 2010 that, according to its 2009 report to the Security Exchange Commission, Exxon

Mobil Corporation (the largest grossing company in the world) paid no U.S. taxes, while reporting a record profit of $45.2 billion. Exxon Mobil minimizes the taxes it pays by using 20 wholly owned subsidiaries in the Bahamas, Bermuda, and the Cayman Islands to legally shelter cash from its operations in Angola, Azerbaijan, and Abu Dhabi (Helman 2010). Exxon-Mobil did pay $17 billion in taxes to other countries, but paid nothing to the United States. While Exxon does not contribute anything to the U.S. federal government, it spends millions on lobbying for the continuation of oil subsidies. According to the Center for Responsive Politics, Exxon Mobil spent over $27 million on lobbying in 2009 alone.

In 2011, as gasoline prices reached $4 per gallon and an angry public prodded Congress, Senate Democrats tried to pass a bill to repeal federal tax breaks and subsidies for five major oil companies. Following several days of hearings during which the Senate criticized "Big Oil" executives, the bill failed 52–48 when three Democrats from oil states voted against it. The bill would have stripped tax breaks from five major oil companies—Exxon Mobil, ConocoPhillips, BP America, Shell Oil, and Chevron. These five companies' profits totaled $35 billion in 2010. Ending their tax breaks would have cost them about $21 billion over 10 years, according to Congress's Joint Economic Committee (Douglas 2011).

The American public has been led to think that it is entitled to cheap energy. Economists generally describe the economics of the 2IR as a conventional or neoclassical economic paradigm that lets the market determine price. However, Hoexter argues, in the case of energy, it is government policies that control the real or final costs (Hoexter 2008). The situation in the EU and Asia offers a telling contrast. In those regions of the world, consumers pay three to four times more than Americans pay for fossil fuels. This creates an economic barrier against overuse and provides an incentive to conserve fuel, seek other options, and use alternative modes of energy.

As the GIR emerges, the world is becoming much more interdependent. What happens in one part of the world, be it weather, pollution, politics, or economics, impacts other regions. For example, the dramatic change in the Egyptian government in early 2010 has affected the rest of the Middle East and will result in global changes of oil and gas supplies. The result might well be the forced end of the 2IR, so that other nations can become energy independent as they implement the GIR.

There is historical precedence for a forced transition from the 2IR to the GIR. The Arab oil embargoes of the early 1970s pushed Europe and Asia toward social policies that eventually led to the development of the GIR.

Energy independence, climate change, and environmental protection became serious political issues. Both of these regions have been developing economic forms of what has become known as "social capitalism," an economic view that includes sustainable growth, environmental concerns, and climate change mitigation, along with interest in diverse populations, gender equality, and democratic processes (Clark and Li 2003). The essence of social capitalism is that there are some social and political problems so complex and so overriding that free markets and deregulation cannot address them.

Social and environmental factors—sustainable communities, climate change mitigation, and environmental protection—are growing in importance and will soon demand far greater international cooperation and agreement. Rampant economic growth and individual accumulation of wealth is being replaced by social and environmental values that benefit the larger community.

Without a national policy, countries cannot address their basic infrastructures and there can be no action, no improvement, no resources, and certainly no response to environmental degradation. For example, the inability of the United States to develop a national energy policy that addresses climate change is often cited as a monumental failure of its free market and deregulation economic model. Energy and infrastructure, the argument goes, are two extraordinarily important national issues. To address them for the greater good, a nation needs to have plans, which are outlined and offered by the central government, to address these basic systems that interact with citizens and the environment.

The key is to have each of the major infrastructure components—energy, water, waste, telecommunications, and transportation—linked and integrated. That way, these components overlap and costs for construction, operations, and maintenance can be contained and reduced. If the basic infrastructure components can be constructed, operated, and maintained on the local level, and meet regional, state, and national goals such as carbon reduction, they take on a different perspective, format, and cost structure. While the United States has no national energy policy, states such as California, New Jersey, Oregon, and New York have created policies of their own in the face of congressional dysfunction and national governmental failure. These days, America's leaders on energy policy are not in Washington, D.C.

The People's Republic of China, not the United States, is showing real global leadership as the world heads into the GIR. More than anything, China demonstrates how important a role the government plays in overseeing, directing, and supporting the economics of technologies and creation of employment. China's economic system is the prototype of social capitalism. Since the 1949 revolution, the Chinese have moved away from communism and toward economic development through a series of five-year plans, now being referred to as guidelines. The central party plays the leading role in establishing the foundations and principles of Chinese policy by mapping strategies for economic development, setting growth targets, and launching reforms. Long-term planning is a key characteristic of centralized social economies, as one overall plan normally contains detailed economic development guidelines for the various regions. Each plan comes with considerable funding to implement the plans based on measureable results.

The Chinese 12th five-year plan addresses rising inequality and sustainable development. It establishes priorities for more equitable wealth distribution, increased domestic consumption, and improved social infrastructure and social safety nets. The plan represents China's efforts to rebalance its economy, shifting emphasis from investment toward consumption and from urban and coastal growth toward rural and inland development. The plan also continues to advocate objectives set out in the 11th five-year plan to enhance environmental protection, which called for a 10 percent reduction in the total discharge of major pollutants in five years.

The current plan will focus the nation on reducing its carbon footprint, as well as addressing climate change and global warming. Not only will it be well financed, with the equivalent of $1 trillion U.S. dollars, but also it will set in motion the possibility that China will be able to surpass Western nations in addressing environmental concerns, creating sustainable communities, and the reaping the benefits of the GIR. Li and Clark (2010), in "Energy Concerns in China's Policy-making Calculations: From Self-Reliance, Market-Dependence to Green Energy," make the case that China has shifted to energy as its key area for political stability, both within the country and for its international policies and programs.

The energy economics for China are considerably different than those for developed countries. For example, one strategy for the Chinese has been to focus on developing nations and their gas and oil supplies. Chinese companies have created joint ventures, merged with, or acquired companies in these regions, causing concern among the developed nations over both the economic and climate change impacts.

Over the years, Chinese economics have changed from a state-controlled communist system to a more socialistic one. In the post-Mao era, China moved aggressively into a "market-capitalism" system, but one where state institutions were owned in part by the Chinese government and shared in joint ventures with foreign companies. Companies wanting to do business in China had to keep their profits there for reinvestment as well as have at least 49 percent of the company owned by the Chinese government. By the end of the late 1990s, China embarked on more of a social-capitalist approach to economic development. Based on the Chinese five-year plans, these finance and capital systems worked extremely well. Today, China has the largest wind (with its partner the Danish company, Vestas) and solar companies in the world. The goal was to create sustainable communities while continuing to grow and expand responsibly to respond to environmental needs and concerns. The environment became an economic asset for the entire nation.

The significant change in China when it leapfrogged into the GIR was its economic growth, which required more secure supplies of fuel and energy. This meant that the Chinese government had to be far more proactive than market economic philosophy would support. While China's economic growth was made possible by an increasing involvement in the capitalist world system, that was also its weakest point. China's economic growth is inseparable from its increasing dependence on global markets, with some estimates suggesting that more than 40 percent of its GNP is derived from international trade (Li and Clark 2010). An area of prime concern is China's energy demand and supply. China's increasing and escalating economic consumption puts pressure on the global energy market supply, affects prices, and causes political and social conflict, especially since China can pay above-market prices for energy.

The question is, can China maintain economic growth in an environmentally sound and responsible manner according to its pledges and commitments? It is apparent that China has identified energy security as one of its vital national interests. It has instituted plans and funds to provide for more energy from renewable sources, rather than rely only on its own limited fossil fuels (or import fuel from other nations). Analysts predict that China will meet or exceed its 2020 renewable energy targets (Li and Clark 2010). China's rise today is due in large part to its rapid emergence as a major force in the world energy markets and energy geo-politics.

Europeans adjusted their economies to fit the requirements of the GIR early on. Both the Scandinavians and the Germans realized that the move away from fossil fuels to renewable energy distribution would require more than the neoclassical free market economics could deliver. While the Danes and the other Scandinavians shifted national resources toward renewable energy power by national consensus, the Germans developed the innovative FiT process.

The German FiT strategy was part of their 2000 Energy Renewable Sources Act, formally called the Act of Granting Priority to Renewable Energy Sources. This remarkable policy was designed to encourage the adoption of renewable energy sources and to help accelerate the move toward "grid parity," making renewable energy for the same price as existing power from the grid. Under a FiT, those generating eligible renewable energy, either homeowners or businesses, would be paid a premium price for the renewable electricity that they produced. Different tariff rates were set for different renewable energy technologies, based on the development costs for each resource. By creating variable-cost-based pricing, the Germans were able to encourage the use of new energy technologies such as wind power, biomass, hydropower, geothermal power, and solar photovoltaic, as well as to support the development of new technologies.

The most significant result of the German FiT was that it stabilized the renewable energy market and reduced the financial risk for energy investment. By guaranteeing investors compensatory payments down to the last pfennig per kWh, the FiT program created a secure climate for investment. The program covered up to 20 years per plant, with the exception of hydroelectricity installations, which required longer amortization periods. The law also offered a means for altering the compensation rates for future installations, if necessary. The executive summary of the original document says,

> This remuneration system does not mean the abandonment of market principles, but only creates the security needed for investment under present market conditions. There is adequate provision to safeguard the future existence of all the plants already in operation. The new act has abolished the regulation contained in the Electricity Feed Act, which limits the uptake at preferential rates of electricity from renewable energy sources to a maximum share of five percent of overall output. Instead, we have introduced a nationwide cost-sharing arrangement. The act should put an end to any fears of excessive financial burdens. The contribution resulting from the new cost-sharing mechanism amounts to a mere 0.1 Pf per kWh. Even if, as we hope, there were powerful growth in renewable energy sources, this would still only rise to 0.2 Pf per kWh in a few years time. That, indeed, is a small price to pay for the development of this key sector. (Federal Ministry for the Environment, Nature Conservation and Nuclear Safety, Germany 2000)

The designers of this policy had exceptional foresight and GIR intuition. The result of the 2000 Energy Renewable Sources Act was the creation of

Germany's renewable energy industry. The policy triggered the creation of wind turbine farms and launched the German solar miracle. Despite having an Alaskan-latitude climate, Germany was—for almost a decade—the number one world leader in solar power manufacturing and installation.

The German FiT is a remarkable GIR economic model. By 2005, 10 percent of the nation's electricity came from renewable sources, when only five years earlier that portion was less than 1 percent. By 2005, 70 percent of Germany's electricity supply was supported with FiTs. Germany estimated that the total level of subsidy was about 3 percent of household electricity costs. The FiT rates are lowered each year to encourage more efficient production of renewable energy. By 2008, the annual reductions were 1.5 percent for electricity from wind, 5 percent for electricity from photovoltaic, and 1 percent for electricity from biomass. In 2010 Germany met their goal of 12.5 percent of electricity consumption, thus avoiding the creation of more than 52 million tons of carbon dioxide. They are on track to reach their goal of 20 percent renewable power generation by 2020.

The German GIR economic model is being adopted by other nations that are developing renewable energy sources. Germany has taken the lead globally to end building of any nuclear power plants which has also had an impact on its corporations. Siemens, the largest builder of nuclear power plants with 17 in Germany alone, announced in September 2011 that it would no longer build any nuclear power plants due. According to the *Renewables Global Status Report: 2009 Update*, FiT policies have been enacted in 63 jurisdictions around the world, including Australia, Austria, Belgium, Brazil, Canada, China, Cyprus, the Czech Republic, Denmark, Estonia, France, Germany, Greece, Hungary, Iran, Republic of Ireland, Israel, Italy, the Republic of Korea, Lithuania, Luxembourg, the Netherlands, Portugal, South Africa, Spain, Sweden, Switzerland, Thailand, and Turkey (Renewable Energy Policy Network for the 21st Century 2009). Despite the lack of a national policy in the United States, several states are considering some form of a FiT, and the concept seems to be gaining momentum in China, India, and Mongolia.

Recently, the European Commission and the International Energy Agency, among other groups, had completed various analyses of the FiT policy. Their conclusion was that well-adapted FiT policies are the most efficient and effective support systems for promoting renewable electricity.

The German FiT model continues to be highly successful, and certainly moves the country beyond conventional 2IR economic theory. The idea that ratepayers can get funds from higher rates to purchase renewable energy systems, which then generate power for their own buildings (with the excess sold back to the central power company), is not part of the neoclassic 2IR economic model. This is because economic theory does not consider infrastructures in their calculations. As Hoexter put it in his article, "neoclassical economics has tended to treat infrastructure as either an externality or just another good or service to be bought and sold on a par with other goods and services" (Hoexter 2008).

FiT—California Style

Of all the states in the United States, California is the most far-reaching in terms of the GIR. California was the first state to launch a major energy efficiency effort after two decades of being one of the most conservation-oriented. The initial three-year funding cycle (2006–2008) has evolved into a multibillion-dollar effort that is now in its second cycle (2010–2012). As part of the state's commitment, California's goal is to generate 33 percent of its electricity from renewable sources by 2020. In 2010, to move this effort forward, California's Public Utility Commission (after more than a year of consideration), approved a version of an FiT called a Renewable Auction Mechanism (RAM). While not a classic FiT, the program is intended to drive small to mid-sized renewable energy development. It will require investor-owned utilities to purchase electricity from solar and other renewable energy systems of 1.5MWto 20MW.

While it is too early to tell if the RAM program will be successful, the renewable energy industry—particularly the solar industry—is optimistic. Several industry leaders say that RAM improves the traditional FiT programs because it allows for market-based pricing, while still providing a long-term, stable power agreement for project developers. Other endorsers think that, because it sets an outcome instead of fixing a price, it will help eliminate speculators and keep high-quality developers involved.

Critics say that California's RAM does too little in terms of both financial support and payback to consumers. The renewable energy industry is skeptical, too, saying that the RAM does not go far enough. The results appear to be similar to the 2IR economic model since direct government rules and standards will likely not be very effective. It does not appear that the state will reach its goal of 33 percent renewable energy by 2020.

Paying to Mitigate Climate Change

Climate change is the greatest challenge of our time, yet, addressing it offers the potential for tremendous economic growth. As we embrace the GIR, efforts to mitigate climate change will unleash a wave of new economic development, generating jobs and revitalizing local, regional, and national economies. However, while the GIR nations may have the technologies to jump-start a green energy economy, developing the mechanism to curb climate-changing emissions is another matter.

Climate change stems from a single fact: human beings treat the environment as a free dumping ground. No one has to pay to pollute our shared air. The result has been increasing concentrations of climate-warming gases, a blanket of carbon in the atmosphere that is keeping heat in the earth's atmosphere.

In order to transition to a GIR green economy, a price must be put on climate-changing emissions. Mechanisms have to be developed that make the polluters pay, while guaranteeing that emission-reduction goals are met. Further, a workable system has to be built on three principles: efficiency, effectiveness, and fairness.

In theory it seems straightforward; however, in practice, and in the political world, it is extremely hard to develop international agreement, especially on economic issues. While several schemes are being discussed, the one being promoted the most by Western 2IR countries is "cap and trade." Proponents say that a fair cap-and-trade system must be comprehensive, operate upstream, allow energy to be auctioned, limit the use of offsets, and have built-in protections for consumers. According to a 2009 report from the Sightline Institute (a Seattle-based think tank), even if the system meets those criteria, it must require nations to set responsible limits on climate change emissions and gradually ratchet down those limits over time. It must also harness the power of the marketplace to reduce emissions as smoothly, efficiently, and cost-effectively as possible, allowing the economy to adjust and thrive.

So what does "cap and trade" mean? A "cap" is a legal limit on the quantity of greenhouse gases a nation's economy can emit each year. Over time, the legal limit goes down—the cap gets tighter—until the country hits its targets and achieves a clean-energy economy. The cap serves as a guarantee that a nation reaches its goal. Countries would use energy efficiency standards for vehicles and appliances, smart-growth plans, building codes, transit investments, tax credits for renewable energy, public investment in energy research and development, and utility regulatory reforms to ensure that the goals are met.

"Trade" refers to a legal system that allows companies to swap the ability to emit greenhouse gases among themselves, thus creating a market for pollution "permits" or "allowances." The point of a trading system is to place a price on pollution that is dispersed through the economy, motivating businesses and consumers to find ways to reduce greenhouse gases. By turning the permission to pollute into a commodity that can be bought and sold, everyone up and down the economic ladder gets new opportunities to make and save money. Ideally the trade leverages the flexible power of the marketplace—the mobilized ingenuity of millions of diverse, dispersed, innovative, self-interested people—to help meet climate goals.

Several cap-and-trade climate policy schemes have already been created; one is being partially deployed in Europe, and another is being proposed for California. The European cap-and-trade system began in 2005, with a first phase, three-year learning period. Despite U.S. criticism that it was a failure, the EU's Emission Trading System (ETS) is stable. According to a report from the World Resources Institute (Duggan 2009), the system has improved through the lessons learned when it was first implemented, and European companies are confident that carbon pricing is here to stay.

Critics claim that the trading system has not changed behavior, but there is some evidence, although questionable, that it has (Dole 2010). For example, in 2006 only 15 percent of the companies covered by the ETS were taking the future cost of carbon into account. Point Carbon and other researchers found that a year later about 65 percent of companies in the trading system were making their future investment decisions based on having a carbon price, which is the system's goal (Duggan 2009). Clark (in press) raises serious questions about cap and trade in terms of its track record both in the United States and the European Union.

The trading system, as an economic model, tries to achieve a balance between supply and demand, which is the core of the 2IR economic model. The problem is that this economic model does not change anything—it allows companies to continue to produce carbon emissions, rather than stopping them. Companies that agree to eliminate their carbon emissions at some point in the future can continue to pollute, postponing their commitment for decades (Dole 2010).

The Carbon Tax Center, a New York nonprofit organization, argues for a carbon tax. They regard a carbon tax as superior to a carbon cap-and-trade system, for five fundamental reasons:

- Carbon taxes will lend predictability to energy prices, whereas cap-and-trade systems exacerbate the price volatility that historically has discouraged investments in less carbon-intensive electricity generation, carbon-reducing energy efficiency and carbon-replacing renewable energy.
- Carbon taxes can be implemented much sooner than complex cap-and-trade systems. Because of the urgency of the climate crisis, we do not have the

luxury of waiting while the myriad details of a cap-and-trade system are resolved through lengthy negotiations.

- Carbon taxes are transparent and easily understandable, making them more likely to elicit the necessary public support than an opaque and difficult to understand cap-and-trade system. The co-author of the U.S. Senate cap-and-trade bill, Sen. John Kerry, even told a reporter in September 2009, "I don't know what 'cap and trade' means. I don't think the average American does."
- Carbon taxes can be implemented with far less opportunity for manipulation by special interests, while a cap-and-trade system's complexity opens it to exploitation by special interests and perverse incentives that can undermine public confidence and undercut its effectiveness.
- Carbon tax revenues can be rebated to the public through dividends or tax-shifting, while the costs of cap-and-trade systems are likely to become a hidden tax as dollars flow to market participants, lawyers and consultants. (Carbon Tax Center 2011)

The world's second-largest emissions trading program was created in December 2010 when the California Air Resources Board (CARB) passed a statewide cap-and-trade bill for greenhouse gas emissions. CARB chairperson Mary Nichols described it as the capstone of the state's climate policy, which aims to reduce emissions to 1990 levels by 2020 (Tollefson 2011).

While California's goals are below what is needed to avoid global warming impacts, it moves the United States closer to a national energy policy. The new regulation sets a statewide limit on greenhouse gas emissions from sources responsible for 80 percent of California's total emissions, covering 360 companies and 600 specific facilities in the initial phase of the program, which runs from 2012–2014. During 2015–2020, distributors of transportation fuels, natural gas, and other fuels will be brought on board.

Instead of identifying a specific emissions limit at a facility, the program mandates that companies have enough emission allotments to cover their annual emissions. They can obtain these through the initial allotment period, when the state supplies them for free, or can purchase them at a later date. In addition, 8 percent of a company's emissions can be covered by projects that offset greenhouse gas emissions elsewhere. Each year through 2020, the total number of allowances supplied by the state declines.

Arguments and criticism abound over cap-and-trade systems and their efficacy. Critics point to the big American companies that are creating trading desks to facilitate "carbon credit" trading. The market could be worth trillions of dollars as emissions reduction becomes an international priority, and critics say that this much money will lead to corruption and cheating. Most of these arguments are being pushed aside as Europe and California adopt cap-and-trade systems. Momentum is building for some way to put a price on pollution. Most likely, an internationally regulated carbon credit trading process will be implemented.

The worrisome thing about the cap-and-trade economic model is that it looks and acts like the deregulatory energy system that failed in the early part of the 21st century in California and elsewhere in the United States. But more recently, cap-and-trade is remarkably similar to the real estate market model that lead to the October 2008 global economic collapse. Traders at the time were wildly leveraging, inflating, and then profiting from purchases based on numbers and statistics later proven to be false. The same people who supported the California deregulation program in the late 1990s, which led to the state's energy crisis, are again pushing the cap-and-trade economic system.

The question is how can climate change be stopped and who will pay for it? Most argue that the voters and politicians need to decide; however, few governments are focused on environmental issues and societal concerns. The GIR economics combines social capitalism with standard economic mechanisms that use externalities. In short, GIR economics is neither the form of economics used in the 2IR nor a totally government-controlled form of finance, regulations, and markets.

Green Jobs: The Promise

The Great Recession of 2008 had started to fade amid signs that the world's financial structure had stabilized and the economy was in recovery. Then, the August 2011 mini-crash hit the markets after an enlarged debt demand by the United States government and a lower Standard & Poor credit ranking, which was reenforced by the Federal Reserve Bank keeping its interests rates low until 2013.

Regionally, Asia appears the strongest—the rampant growth in China hardly slowing. The EU and the United States are lagging, for different reasons. Europe is trying to maintain a unified structure while balancing the wealth and stability of the more prosperous nations such as Germany, France, and the Nordic countries, against the near bankrupt smaller nations such as Greece, Ireland, Spain, and the 25 new central and eastern EU member nations that are rapidly trying to catch up.

The U.S. economy has a slow recovery ahead, given its near-complete collapse. Massive government debt and a stubbornly high unemployment rate are hindering this process. Unlike the case with past recessions in the

United States, this economy is recovering with the help of America's finance and technology sector, not springing back due to re-employment and new job creation. This top-down recovery is difficult, as fewer people benefit and the process is much slower. The U.S. stimulus package seems to have benefitted the financial and investment markets rather than stimulated job growth. In a finance and technology boom, wealth is disproportionately shifted to a smaller number of people, making a jobless recovery that much more painful for most citizens.

In the United States, the job market has been slow to respond to the Federal Reserve stimulus package because the economy lost the hugely leveraged housing industry, a major economic driver and a sector that employed large numbers of unskilled and low-skilled workers. Since the 1980s, the U.S. manufacturing sector, which had been the source of high-paying union jobs, has been in steady decline. A high-tech knowledge-based economy has emerged to replace the manufacturing economy, but this type of economy skews toward those with a college education and technical and financial acumen. Once the housing market collapsed, hundreds of thousands of blue-collar workers were shut out of jobs that may never come back, just as manufacturing workers of the 1980s and 1990s were displaced from factory jobs when mechanical engineering moved rapidly into electrical engineering, signaling the end of the 2IR.

The United States, unlike China, Japan, South Korea, and Europe, has been slow to embrace the enormous potential for job creation in the Green Industrial Revolution. While the current presidential administration identifies "green" job growth as the way to reduce the unemployment level, the U.S. Congress seems locked in fossil fuel–driven 2IR economics and its associated public policies.

The Chinese have leapfrogged the rest of the world into GIR economics, technology, and job creation. In fact, China is now leading the world with economic and career innovations for the GIR. Analyzing China's phenomenal economic growth, it is clear how much of that growth can be attributed to the development of green industries associated with the GIR. Not only is China creating massive systems to generate renewable energy for its own use, but also it has quickly become the world leader in exporting these technologies. If anyone around the world wants to buy solar panels, or wind turbines, China is able to provide the best pricing and quality.

What the Chinese clearly understand and other nations do not is that once committed to the GIR, a nation creates new economic development and business opportunities that lead directly to job generation, new career paths, and the revitalization of local economies. This results in further research and development supporting the GIR. Because of the extraordinary interconnectedness of a modern nation's economy, once local economies start to come back, they revitalize the service industries that support them—the markets, retail stores, and small businesses, as well as schools, city governments, and all the other public agencies dependent on tax revenues.

So what exactly is a "green" job? The label is much like the "knowledge-based" job label, a generic term that describes an industry or service rather than a specific type of activity. Raquel Pinderhughes, a professor of urban studies at San Francisco State University, defines "green jobs" as a generic term for people doing any kind of work, whether mental or manual, that in some way relates to improvements in environmental quality (Pinderhughes 2006). Dr. Anthony "Van" Jones, a community activist in Oakland, California, takes the concept further by asserting that green jobs should be the only kind of jobs and careers that should be supported if society is to reverse the impacts of climate change (Friedman 2007).

The UN Environment Programme's *Green Jobs: Towards Decent Work in a Sustainable Low-Carbon World,* added a subset called "green-collar" jobs. The UN tried for a more rigorous definition of green jobs, saying, "This includes jobs that help to protect ecosystems and biodiversity; reduce energy, materials, and water consumption through high efficiency strategies; de-carbonize the economy; and minimize or altogether avoid generation of all forms of waste and pollution" (UN Environment Programme 2008). A Silicon Valley nonprofit organization named Next 10 was founded to not only track changes in climate change through reporting on public policy and data, but also to provide statistics on the kinds and numbers of new green jobs that are created in California.

Pinderhughes says that green-collar jobs are the specific manual labor opportunities in a green economy that would be open to low-skilled workers in industries, such as bicycle repair, recycling collection, and waste composting. These are entry-level positions for which people can be quickly trained. Pinderhughes, whose research was commissioned by the Roberts Enterprise Development Fund, a San Francisco–based philanthropy fund, thinks that the green economy could provide a pathway out of poverty for people who have been considered chronically unemployable. Existing green-collar jobs in the Bay Area, her research has found, offer living wages, good working conditions, and occupational mobility, advantages that typically don't come with traditional blue-collar work (Pinderhughes 2006).

It is difficult to put a number on how many jobs would be created if the U.S. economy focused on climate change mitigation. Economists struggle when they analyze green job data and try to interpret the results. Robert Pollin, co-director of the Political Economy Research Institute at the University of Massachusetts, wrote a report in 2008 that calculated the United States could generate two million new jobs over the next two years with a $100 billion investment in a green recovery (Pollin et al. 2008). President Obama talked about creating five million new green jobs in his 2011 State of the Union address. As with any emerging industry, the actual number of jobs that will be created is hard to predict.

While economists can provide calculations that show on average how many jobs would be created, based on the number of dollars invested, they cannot measure the corollary impact or predict the number of related jobs that will be created. However, one precedent for green job creation is Germany's FiT. Germany attributes strong growth in the renewable energy sector to blunting the recession. According to Deputy Environment Minister Astrid Klug, there were now 250,000 jobs in Germany's renewable energies sector and an overall total of 1.8 million in environmental protection. The number of jobs in renewables will triple by 2020, and hit 900,000 by 2030.

Another example is California's world-leading energy efficiency program, which has put thousands of workers back to work retrofitting commercial buildings. At the same time, the program is driving the lighting industry to develop extraordinary new products (like decorative, efficient, and cost-effective LED lights) and demand responsive dimmable ballasts. This new generation of lighting products is transforming the market and taking it into the GIR, along the way creating a potential $1 billion industry for California. If LED and dimmable ballast changeovers were made a national priority, it would be a multibillion-dollar industry in the United States alone.

Private Investment Is Lining Up

Goldman Sachs, the world's premier investment bank, noted on its website that harnessing an important natural resource like wind requires a substantial amount of capital. That is why they helped renewable energy companies find the capital they needed to embrace this innovative technology. They go on to note that they have invested $3 billion in clean energy since 2006, and raised more than $10 billion in financing for clean energy clients around the world since 2006 (Goldman Sachs 2009). The key factor, however, when looking at any investment from venture capital, corporate, banking, or private equity companies, is to keep in mind that there is a difference between "clean" and "green" technologies. Google has recognized that difference. As reported in July 2011, Google has invested more than $100 million in off-shore wind turbines off the U.S. east coast.

If Goldman Sachs, one of the icons of 2IR economy, and Google, one of the leaders in the GIR, are both prominently promoting investments in the GIR, it must mean that venture capital investors are starting to pay close attention. Venture Business Research (VB/Research), a leading data provider, tracks the international financial activity in the clean industry. They reported in the second quarter of 2010 that venture capital and private equity investment in clean technology and renewable energy exceeded $5 billion worldwide, despite a 30 percent decline in early stage venture capital activity. They also reported that a record number of merger and acquisition deals valued at over $14.5 billion were transacted during that period. They noted that the initial public offering sector was dominated by Chinese companies, which accounted for 75 percent of new issues, and stated, "The market lacks pre-credit crunch exuberance but has recovered significantly from the moribund levels of 2009" (Venture Business Research Limited 2010). Then they noted that the $5 billion invested during the second quarter of 2010 in global venture capital and private equity funds was slightly ahead of average quarterly investment of $4.9 billion since the second quarter of 2008 and 45 percent above the low 2009 level.

The bulk of the money being invested in green/clean technology is going to China, according to accounting firm Ernst & Young. The firm does a quarterly assessment of the most attractive countries for renewable energy investment. In their September 2010 report, they noted that China was now the world's biggest energy consumer, as well as the world's most attractive country for renewable energy investment (LaMonica 2010). China has been encouraging investment in its clean energy companies as part of its goal of generating 15 percent of its electricity from renewable sources by 2020. Ernst & Young also compared regulations, access to capital, land availability, planning barriers, subsidies, and access to the power grid. The report ranked investments in onshore/offshore wind, solar, biomass, and geothermal energy projects. After China, the next most attractive countries for renewable energy investment were the United States, Germany, India, and Italy. Ernst & Young noted that government support in China gives it a huge advantage over other countries in pursuing clean energy projects. In the second half of 2009, China almost doubled consumer subsidies for generating renewable power, bringing the amount to $545 million (LaMonica 2010). The fact that China has both five-year plans and substantial financial support for the green technology sector gives it a huge advantage in the GIR. Even more significantly, the Chinese require government participation or partial ownership of many of the new firms, in keeping with their tradition of government control of infrastructures.

The Ernst & Young report underscores the extraordinary progress China has been making as it moves into the GIR. Clearly, China recognizes the enormous potential for economic growth that can occur through a concentrated national policy to mitigate climate change. Over the last decade they have pushed past the United States and Europe in renewable technologies, business development, green job generation, and economic revitalization. Not only has China taken the lead in renewable energy technologies, they have been far advanced in developing a national planning process that sets environmental protection as a priority. Now, as Ernst & Young's report makes clear, they are capturing the world's investment capital.

Creating an economy that can move the world into the GIR is an exceptionally complex process. Various nations and regions are approaching the problem differently. The European FiT program and China's direct government subsidies have been the most successful efforts. Some U.S. states, such as California with its newly designed Renewable Auction Mechanism (RAM), have developed possible improvements over the European FiT. But the RAM is much more limited and available only in California.

So the Green Industrial Revolution is not all about the money. It is about climate change mitigation, renewable energy, smart grids, and environmental sensitivity. But achieving the benefits of the GIR—a wave of new technologies, business enterprises, and green jobs—will require substantial public and private financing. A new green GIR economy will be needed to accelerate the necessary changes and stop climate change.

Chapter 11

The Race for Global Energy Innovation Has Begun: Where Is America?

America continues to support the companies and technologies of the carbon-dependent economy that created a Second Industrial Revolution, despite the cost. Exxon Mobil made $45 billion in profits in 2009, yet paid no U.S. federal income tax (Helman 2010). Meanwhile, the rest of America suffered through the worse economic decline since the Great Depression of the 1930s. General Electric (GE) earned $10.3 billion in pretax income the same year but also paid nothing to the IRS (Helman 2010).

Loss of tax revenues are just part of the story. *Time* magazine reported in February 2011 that Chevron intends to continue to fight a 17-year-long court case that arose after its 2001 merger with Texaco. An Ecuadoran court ordered Chevron to pay $8.6 billion to clean up oil pollution in the country's rain forest. According to *Time*, the court cited the oil giant because Texaco allegedly dumped "1.8 billion gallons of toxic wastewater and spilled 17 million gallons of crude oil that caused an estimated $27 billion in damages over three decades and triggered health problems like cancer and skin disease" (Padgett and Kuffner 2011). Chevron says that it will not abide by the ruling.

Meanwhile in early 2011, the Middle East experienced another kind of revolution: a human one through social media. Revolutions and changes in leadership—first in Tunisia, next in Egypt, then in Libya, now in Syria, and soon perhaps in other Arab nations—have created uncertainty. This may be the beginning of the end of cheap fossil fuel from the Middle East. On the positive side, these revolutions could be a catalyst to jump-start America's entry into the GIR, and to push the United States toward renewable energy independence.

This new era of sustainability and carbonless energy generation is at America's doorstep. The Japanese and South Koreans have been leading the effort since the 1970s, and Europe has been involved for the last two decades. This push for renewable energy will become history's largest social and economic megatrend, providing extraordinary benefits in the form of economic revival, innovation, emerging technologies, and significant job growth for those nations capable of fast entry. Depending on policy and some critical short-term decisions, this may be the first major megatrend since the initial Industrial Revolution that won't be led by American inventiveness or know-how.

America has been at a standstill now for over a decade, thrown off course by 9/11 and a century-old 2IR, with its dependence on cheap fossil fuels, antiquated grid systems, and collapsing infrastructures. In the post-9/11 era, America exported its fears and disconnected itself from many of its allies. The United States has slipped into bad habits that have weakened the nation's ability and willingness to take on the bigger challenges. As the 10th anniversary of 9/11 came, America had compromised its moral leadership, technological innovation, and sustainable growth through a series of poor decisions and the financially devastating "credit swap" and subprime mortgage crisis. America's careless actions have led many former friends and allies to the brink of economic chaos and social discord.

The nation that gave us the heroes of World War II has fallen into a defensive crouch. Worse, it has fallen into the politics of "no" or, as described by *New York Times* columnist Thomas Friedman, the politics of "as dumb as we wanna be" (Friedman 2008).

This political dysfunction, with its constant demonizing of political opponents and concentration on the trivial at the expense of the important, has forced America to cling to the former glory of the Second Industrial Revolution when the rest of the world wants to get on with cleaning up the environment, reducing greenhouse gases, and responding to global warming.

Why is America still locked into coal as its main source of energy? Why do national leaders allow legal loopholes for companies that "frack" for natural gas, seek shale oil at any cost to the environment, continue to drill off-shore wells, and pour toxins into water supplies? Where are the new tools, like the FiT, to promote renewable energy generation in America?

In many cases, the gas and oil companies are driving the policies. Even in California (the nation's most progressive state when it comes to energy efficiency, energy renewables, and environmental issues), there was an initiative on the 2010 ballot to roll back air quality standards for the benefit of the big fossil fuel companies. The large oil companies had sponsored the ballot measure (Roth 2010). Fortunately, it failed by a significant majority.

What happened to America's values of hard work, achievement, and accountability? The country will need to rediscover these core values if Americans are to enjoy the benefits of the carbon-less economy in the global Green Industrial Revolution.

Jeremy Rifkin's concept of the Third Industrial Revolution (3IR), introduced in his 2004 work *The European Dream*, was prescient. Although Rifkin saw the 3IR first in Europe, it actually began in Japan and South Korea more than two decades earlier. After the 1973–1974 Arab oil embargo, the Japanese—as well as some Europeans—realized that there was an overwhelming political need to pursue energy independence. They responded by raising gasoline taxes and launching energy efficiency efforts. Yet the historical traditions, cultures, and needs of Japan, for example, have always been that as an island nation, it had to have supplies of energy from within its borders or shipped in. So for many decades, especially since the end of World War II and now with the aftermath of the earthquake and destruction of the Fukushima nuclear power plant, the Japanese have used their cultural imperative to create profitable businesses that conserve energy and water while also recycling waste and raw materials.

Slowly, countries around the world are realizing the need to replace fossil fuels with renewable energy generation. France invested heavily in nuclear energy and today France gets more than 70 percent of its energy from nuclear power ("France: Energy Profile" 2007). Nuclear power is not France's best alternative, but one that will buy more time to develop renewable energy sources. Even a developing country, Brazil, got into the act by launching a national program to generate energy from ethanol made from the country's abundant sugar cane resources. Now, with a combination of domestic oil and ethanol, Brazil is energy independent. These countries realize that their futures are not rooted in the Second Industrial Revolution and they are finding better energy solutions in the GIR.

For a while, America was on the right track. It was Republican president Richard Nixon who signed the nation's first major environmental laws, which addressed air pollution, water pollution, and toxic waste. Under presidents Gerald Ford (Republican) and Jimmy Carter (Democrat), the United States raised mileage standards for cars and trucks. Congress even passed the 1975 Energy Policy and Conservation Act (CAFE), which required the gradual doubling of passenger car fuel efficiency to 27.5 miles per gallon. To the astonishment of no one, it worked, and between 1975 and 1985, the average American passenger vehicle mileage went from 13.5 to 27.5 miles per gallon. Mileage averages for light trucks went from 11.6 to 19.5 miles per gallon (Union of Concerned Scientists 2007). These efforts contributed to the global oil glut of the late 1980s, the eventual weakening of OPEC, and the unraveling of the Soviet Union, which at the time was the second largest oil producer.

Unfortunately, instead of continuing these forward-thinking policies, President Reagan slashed CAFE, and ended the alternative energy programs started by President Carter. Carter's vision of an energy independent America went up in fossil-fuel smoke. President Reagan and Congress let the tax incentives for solar and wind start-ups lapse. Ironically, several of these alternative energy companies, which had been funded by American taxpayers, ended up being bought by Japanese and European companies. As a result, American technology helped propel those countries into the GIR. Reagan even stripped off the solar panels that Carter had mounted on the White House as a symbol of alternative renewable energy generation.

Reagan not only destroyed America's nascent alternative energy industry, but he also ended the bipartisan approach to the environment. He and James Watt, his secretary of the interior, systematically turned America's historic respect for the environment and natural world into a partisan and polarizing issue.

Reagan was followed by Presidents George H. W. Bush, William Clinton, and then George W. Bush, all of whom ignored the need to tighten fuel mileage standards. In 2003 the Chinese leaped ahead of the United States in fuel standards. Finally, in late 2007, Congress reinstated the 27.5 miles per gallon standard and pushed future standards to 35 by 2020. Both Europe and Japan already have standards of 35 miles per gallon. Now, decades later than would have been the case, Americans are starting to grasp the fact that big oil and big auto are not particularly concerned with the nation's well-being.

America's Post-9/11 Funk

Except for some visionary academics and some prescient venture capitalists who were hard at work raising funds to invest in the emerging green industry, few Americans saw this new industrial trend coming.

The years leading up to September 11, 2001—the Clinton years—were a time of American splendor. Money was everywhere, fueled by the information technology bubble, then the Internet bubble, then runaway housing prices. It was the time of "cash-out" and easy refinancing.

Then on 9/11, America's world vision and confidence were rocked. After the horrific events of that day, America was afraid. The nation reacted in the worst possible way, led by a president who was more focused on retribution than the causes of 9/11. Instead of concluding, as the Europeans and Japanese did in the 1980s, that energy independence was the key to security, America went after the Middle East with a vengeance. America forgot what history makes clear: in the end, the Sahara's sands will wipe away all conquerors.

There was a pivotal moment when America could have led the push for renewable energy. The moment was on September 14, 2001, when President Bush stood at Ground Zero, in New York City. Had the president called for a future without foreign oil and asked for the national sacrifice needed to make the shift to energy renewables a reality, America would now be awash in environmentally friendly energy as well as being the international leader of the GIR. Imagine the enormous number of well-paying jobs that would have been created had America embraced the carbonless economy in 2001. Instead, this Texas oil politician took the road the nation should have never traveled, and, once again, America embraced fossil fuel and its right to use a disproportionate amount of the world's oil supply to support a lifestyle that is not suited for an overpopulated planet plagued by climate change.

It is not too late, which is the message of this book. There is still time for America to do what it does best—take on a challenge, dig in, and get the job done. The opportunity still exists to leapfrog China and take the leadership of the GIR. While Congress and the rest of the federal government continue to drift sideways, there is time to embrace the GIR's massive transformation of resources away from a lifestyle powered by fossil-fuel and carbon-based energy to one that uses renewable energy distributed by smart grids, under a system that is community-based, flexible, and environmentally friendly.

America desperately needs restoration and revitalization. There is still time for the country's innovators and idealists to push forward their ideas about making clean energy. Many of these ideas are cutting-edge, and the large number of experimenters shows that the innovative spirit that drove the creative breakthroughs of Hewlett and Packard is still alive in the nation's garages. America needs a massive national effort to find game-changing solutions, as occurred during the economic and life-changing explosion of the information technology era.

If America accepts the challenge and pursues GIR leadership, it must push forward the new "green" economy at the expense of the old "dirty" economy. America is already starting to move from an economy based on extraction, natural resources, and fossil fuels to one based on knowledge and renewable energy. Momentum is developing, pushed by necessity.

Most of the progress has been at the state level. California is in the second cycle of a multi-billion dollar energy efficiency program. The program is about to launch a $3.2 billion home retrofit program that will employ hundreds, if not thousands, of struggling construction workers. California also has a growing commitment to renewable energy and Governor Jerry Brown pledges to continue to push these initiatives. New York has launched a major energy efficiency program just in time. Manhattan, one of the world's most dynamic and energy-intensive environments, has been teetering on the brink of grid capacity. At this point, if New York wants to avoid energy disruptions, the city has to reduce demand and increase energy efficiency as the first step. These pressures, combined with the need to develop new jobs (most likely "green" jobs) for an economy with chronically high unemployment, will provide multiple drivers for the new GIR economy. Each step opens the door a little more.

The initial Industrial Revolution was a turning point in human history. Great Britain led the transition from a manual labor, draft animal–based economy to machine-based manufacturing. The Age of Enlightenment offered new and startling ideas, and pushed art, literature, science, and democracy into new frontiers. This revolution was built on coal and steam, with Watt's steam engine powering textile and manufacturing industries that were the envy of the world.

The Second Industrial Revolution started with the nexus of the fossil fuel–powered internal combustion engine and the beginnings of telephonic and electronic communications. The internal combustion engine, along with the commercialization of oil, created a previously unimaginable world of machines and popularized personal transportation. The telephone revolutionized the daily lives of ordinary people.

The third and green industrial revolution is more significant and life changing than either of the previous industrial revolutions, for there is so much more at stake. The world has become very crowded; each day resources get scarcer. The UN predicts that by 2053, there will be 9 billion people living on the planet (UN Department of Economic and Social Affairs 2011). Today there are 6.9 billion.

Compounding the problem is the rise of a middle class in developing nations. Large groups of people in places such as India and China want out of poverty, and who can blame them? They want the things that developed nations already have—nice clothes, nutritious food (including animal protein for their kids), and large, air-conditioned, electrified homes. They also want the things that most Americans consider basic necessities: washing machines, cell phones, refrigerators, televisions, and (most environmentally damaging) cars. The universal desire for the world's developing classes is a car—the epitome of personal transportation, wealth, and status.

Add it up, and the world will soon be resource-constricted, particularly since the planet is running out of fossil fuel. This alone threatens to shake the very foundation of our existence. But the environmental degradation and in some cases collapse of various parts of our planet's ecosystem is adding a heightened sense of urgency.

The Earth is getting hotter, and climate change is real. The planet is being threatened by melting ice caps, ocean acidification, loss of plant and animal species, and extreme weather patterns that impact sensitive environments.

A Dangerous, But Opportune New Age

We are entering a dangerous age—a point at which global warming and environmental degradation may become irreversible. Critical decisions must be made on a global level for the good of the planet. It is also an age of opportunity, and the Green Industrial Revolution will provide those opportunities.

At the foundation of this new era will be sustainable and smart agile communities. Europe and Japan have set the pace for sustainable and energy-secure communities, through the use of their own renewable energy sources, storage devices, and emerging technologies. Nations, states, and cities want to control through centralized power and authority, which has been the historical pattern. However, with the need to mitigate climate change, regional and local solutions must be developed.

A sustainable smart community is part of a living network that draws residents from a broader region. Agile, sustainable communities must develop environmentally friendly ways to generate renewable energy, handle waste, conserve water, and develop rational transportation and telecommunication systems. Sustainable communities must transition from the traditional central power plants and unsustainable infrastructure systems to ones that use onsite renewable energy, recycling, waste control, water supplies, green building standards, and land management systems.

Renewable energy, distributed by smart green grids, will help create sustainable communities of all sizes, shapes, cultures, and languages. Systems that use regenerative brakes, flywheels, or fuel cells to provide practical and efficient energy storage will assure consistent power generation. Smart grids will constantly monitor and redirect power to where it is needed in homes, office buildings, schools, and transportation systems.

Most American cities have the potential to implement some, if not all, of these activities. With a little guidance, our communities can have locally distributed renewable energy, clean water, recycled garbage and waste, and efficient transportation systems. Developing dense, compact, walkable communities that enable a range of transportation choices conserves energy and makes economic sense. America must become sustainable and free itself from the carbon-intensive, fossil fuel–based, inefficient, centralized energy generation systems of the 2IR. Instead of lagging behind, America needs to become the leader in a world on the cusp of historic change.

Agile, sustainable, and smart communities are a necessity for a less-polluted environment today and a green world tomorrow. The solutions to global warming and climate change exist now; America needs to design and implement them. Joining this new carbonless economy will not only save America millions of dollars and millions of tons of toxic greenhouse gases, it is also the key to obtaining energy independence from foreign oil and gas.

As the leader of democracy for two centuries, America must examine its core values and provide humanity's future direction. American leadership needs to re-establish a global vision with strategies that move rapidly into the Green Industrial Revolution, an era that is no longer dependent on the fossil fuels or nuclear power that defined the Second Industrial Revolution, but one that generates stationary power and creates fuel from renewable energy sources. America must regain the technological inventiveness and entrepreneurship that drove the high-tech boom of the late 20th century.

Equally as important, America needs to reduce its energy dependency on the Mideast, a geopolitical region in turmoil and the channel for the most massive wealth transfer of all time—from the world's oil consumers to the oil suppliers. This transfer of wealth is creating political instability that threatens America's national security, drains precious financial resources, and keeps the nation from focusing on crucial domestic issues.

Since 2005, when Hurricane Katrina destroyed much of New Orleans, climate change and global warming has been a reality for America. Now, with clear evidence that global warming and climate change is impacting our daily lives, the nation needs to quickly convert to renewable energy and a sustainable future.

There are simple, easy ways to slide into this new era. For example, the current economic downturn has states reeling with the loss of tax and real estate development revenues. Much of this downturn is rooted in the 2IR, and while Europe and Asia see that fossil fuels are not their economic future, the United States continues to invest in them; supporting offshore drilling, mining, and scalping mountains while polluting rivers and oceans. A simple step would be to tax the 2IR fuels and then invest those revenues in renewable fuels and smart technologies.

Other simple steps could make huge impacts. One would be to establish a national energy policy that puts renewable energy generation at the forefront, like Chile did at the end of 2010 and Germany did again with its reaffirmation of a FiT to fund its national plans. Another would be to require all new cars sold in America to be hybrids, and all electric-powered vehicles to use energy from renewable power sources. Building codes should require that new houses and commercial facilities be built with solar panels and simple biomass energy generating systems that harvest organic waste.

Even simpler would be to require the nation's utilities to switch from coal and fossil fuels to renewable energy (traditionally, American utilities control the source of energy generation). Unfortunately, the utilities would no doubt prefer the more profitable route of establishing their own huge wind and solar farms, even though it's much easier and more environmentally sensitive to use existing roof structures for solar panels. A few efficient solar panels on every rooftop is not what a utility wants, yet this sort of decentralization of energy generation is the key to the GIR. The use of a carbon tax (or incentive to fund renewables) would support this change. Thus, the nation's large utilities will have to undergo extraordinary changes as the GIR takes hold.

Whether America is ready or not, the era of the Green Industrial Revolution with its global energy innovation is at its doorstep. The place to start is with small, relatively self-contained communities or villages within larger cities and regions. America must get these communities off their dependency on central grid–connected fossil fuels and make them sustainable from renewable power sources that do not harm the environment.

Renewable onsite power is more efficient when locally used, similar to the way communities in Denmark are generating power with a combination of wind and biomass to provide base load (Lund and Ostergaard 2009). The goal must be sustainable communities linked to sustainable states and eventually to a nation free of oil dependency from foreign and domestic sources. In return for developing a carbonless GIR economy, America would get a new national focus that would reinvigorate and renew its national pride, strengthen its international position, and help it retain its wealth instead of transferring it into the bank accounts of petro-dictators. In addition, the effort of developing a new energy generation system will provide vast investment opportunities and create the green jobs needed to replenish the tax revenue and aid the nation's failing cities and educational systems.

For America, energy independence and the subsequent elimination of energy bills with foreign postmarks are two of the potential benefits of entering the GIR. As soon as possible, America needs to give up freebasing fossil fuels and embrace a healthier lifestyle with intelligent development and greater community connectivity. What is crucial is that Americans, starting in local communities, see the vision and take action. Almost every community has the renewable resources to make itself energy independent and carbon neutral. America must get started, develop a funded national energy policy, and then get out of the way and let the nation's historic talents for innovation and entrepreneurship take over.

While citizens and communities can do much, the U.S. government needs to shake itself out of its existential dysfunctional behavior and start acting like an institution of leadership. Fossil fuels were not cheap when they first came into use in the late 1890s as the foundation for the Second Industrial Revolution. An initial taxpayer investment created the big centralized fossil-based power plants and large-scale central grids necessary to develop and deliver power to far-flung communities.

A series of laws at the turn of the 20th century allowed power suppliers to turn into economic monopolies. Initial government support and incentives evolved into a century-long set of subsidies for big oil and coal, some of which still exist. As a result, these large fossil-fuel suppliers, power generators, and resource extractors remained America's dominant economic business organizations until the end of the 20th century.

Now that the price of oil hovers around $100 a barrel and oil companies are among the richest companies in the world, why are they still allowed special incentives and tax breaks? It's time to stop, especially as America tries to balance its budget. The incentives of the 2IR for fossil fuel must be reduced and applied to renewable energy generation, which is the basis of the GIR. This type of tax shift has been successful in other nations, without adding to the taxpayers' burden. Renewable energy technologies and their integrated systems deserve the same kind of policy and financing that fossil fuels got.

To maximize efficiency, renewable energy generating systems need to be integrated as linked or bundled supply sources based on the natural physical characteristics of the location. Further, these intermittent power-generating resources can be greatly enhanced by using storage devices, because the sun is not always shining and the wind is not always blowing.

Energy independence will not be achieved overnight, any more than the SUV and the carbon-intensive economy became social and political realities instantly. Americans have spent a trillion dollars on the wars in Iraq and Afghanistan, and it will probably cost at least that much to turn America into an energy independent, sustainable nation of the GIR. However, concerns over national survival, global warming, and international political leadership are compelling us to quickly surpass what has begun in Europe, Japan, South Korea, India, and China.

While energy efficiency is a first and important step, complete energy independence and carbon-neutral policies and programs are within America's technological grasp. Another generation of renewable technologies is coming and it is much better—lighter, thinner, stronger, and cheaper. Wind and solar power, coupled with highly efficient storage devices, smart green grids, and local onsite distribution systems are within reach.

A new global economy is coming. It includes a dramatic shift from the neoclassical economics of energy (fossil fuels and nuclear power) to the new economics of societal concerns ranging from renewable energy (green technologies) to public health care. These technologies are becoming more and more commercially attractive and cost-effective, and consumers are beginning to install solar and wind energy systems at their homes, offices, and farms. What America lacks is the large national financing and political leadership to make the commitment and lead the nation into energy innovation on a national scale.

Each nation needs to have an overall master plan to address the basic systems that interact with its citizens and the environment. That plan should cover how government, business, and the public address basic infrastructures such as energy, water, waste, telecommunications, and transportation. Without a national plan, there is no action, no improvement, no new resources, and certainly no response to environmental degradation.

One country that has taken this approach is China. Since 1949, China has had five-year plans that provide the government and population with direction from the central government about the policies and programs that need to be pursued. More recently, these five-year plans have included business development in what the Chinese refer to as social capitalism, which means that business is important, but it must consider the concerns and values of the people.

National plans need to link and integrate public infrastructure components. That way, infrastructures overlap and costs for construction, operations, and maintenance can be contained and reduced. If infrastructures can be constructed, operated, and maintained on the local level and meet regional, state, and national goals such as carbon reduction, they take on a different perspective, format, and cost structure. Economics is the key.

In some of the emerging GIR countries the central grid is being replaced, as onsite power generation takes the place of large fossil fuel or nuclear power plants. These updated grids allow reverse power flow, which captures surplus energy generated from residential solar installations. This is a necessity because when hundreds of small residential systems are concentrated, the demand can easily strain existing power distribution systems. For example, on a sunny day, the surplus power generated through solar energy could easily exceed the upper limit for the voltage management system. The new smart green grid systems are able to manage such variations.

America can no longer continue to deny that it needs to take a new path. The nation must come up with a national energy policy that makes sense. The entire country must move rapidly from the fossil fuels that dominated the 20th century, to the renewable energy GIR era, which will be the new world order of the 21st and 22nd centuries. The nation needs to catch up, and the sooner it starts, the faster it can achieve the benefits of a healthier sustainable lifestyle.

The solution is for America to retake its place as world leader. America led the second industrial era, and by the end of the 20th century, it was the world leader in innovation and entrepreneurship. By the Millennium (21st century), America was creating historic advances in computerization and was the world leader in information technology. Now that distinction as innovator and entrepreneurial dynamo is challenged, as the world seeks leadership in the most crucial battle of all: to stop global warming and reverse climate change. Now is the time for America to wave good-bye to fossil fuels and step up as world leader of the new carbonless Green Industrial Revolutionary era.

References

American Council on Renewable Energy (ACORE). 2011. *US–China: Quarterly Market Review*, Spring. http://www.acore.org/wp-content/uploads/2011/05/ACORE_US_CHINA_QMR-spr2011_ES_FNLhi.pdf.

American Wind Energy Association. 2011. "U.S. Wind Industry Fast Facts." http://www.awea.org/learnabout/industry_stats/index.cfm.

APCO Worldwide. 2010. "China's 12th Five-Year Plan: How It Actually Works and What's in Store for the Next Five Years." APCO Worldwide. http://www.apcoworldwide.com/content/pdfs/chinas_12th_five-year_plan.pdf.

Arctic Council and the International Arctic Science Committee (IASC) Secretariet. 2004. Tromsø, Norway.

Arctic Monitoring and Assessment Program. 2009. "Update on Selected Climate Issues of Concern." http://www.amap.no/.

Asola, Teresio, and Alex Riolfo. 2009. "Sustainable Communities: The Piedmont Region, Settimo." In *Sustainable Communities*, edited by Woodrow W. Clark II, 169–92. New York: Springer.

Blumer, Herbert. 1986[1969]. *Symbolic Interaction: Perspective and Method.* Englewood Cliffs, NJ: Prentice-Hall.

Boudway, Ira. 2011. "Bloom Energy's Fuel Cell Breakthrough." *Bloomberg Businessweek,* January 20. http://www.businessweek.com/technology/content/jan2011/tc20110120_212633.htm.

Bradsher, Keith. 2010. "China Leading Global Race to Make Clean Energy." *New York Times,* January 30. http://www.nytimes.com/2010/01/31/business/energy-environment/31renew.html.

Brian, Marshall. 2000. "How Power Grids Work," April 1. Northhampton, MA: Clark Science Center, Smith College. http://www.science.smith.edu/~jcardell/Courses/EGR220/ElecPwr_HSW.html.

British Petroleum. 2011. "China the Fuel for Growth," Issue 1. http://www.bp.com/sectiongenericarticle.do?categoryId=9037009&contentId=7068199.

Brodrick, J. James. 2006. "DOE Solid State Lighting Status and Future." Report for the Department of Energy. http://apps1.eere.energy.gov/buildings/publications/pdfs/ssl/brodrick.pdf.

Brown, Lester. 2009. *Plan B 4.0 Mobilizing to Save Civilization*. Earth Policy Institute. http://www.earth-policy.org/images/uploads/book_files/pb4book.pdf.

Brundtland, Gro Harlem. 1987. *Our Common Future: The World Commission on Environment and Development.* New York: Oxford University Press.

California State Law, Assembly Bill No. 1109. 2007. Sacramento, CA, October 12. www.climatechange.ca.gov/.../ab_1109_bill_20071012_chaptered.pdf.

Callahan, David. 2010. *Fortunes of Change: The Rise of the Liberal Rich and the Remaking of America.* Hoboken, NJ: Wiley.

Carbon Tax Center. 2011. "Introduction," January 31. http://www.carbontax.org/introduction/.

"Car Ownership in China Expected to Overtake Japan Next Year." 2010. *People's Daily Online,* May 31. http://english.peopledaily.com.cn/90001/90778/90860/7006415.html.

Chan, Steven. 2011. "Global Solar Industry Prospects in 2011." SolarTech. Summit, Santa Clara, California, March.

City of Benicia, California. 2009. "2007–2009 Bi-Annual Budget." www.ci.benicia.ca.us/index.asp?Type=B_BASIC&SEC={58DD8C47–2DC8–4491-B07E-FCBDF4C3B17E}&DE={D9CC3122-E111–4DB7-BDAC-658307F0AEB1}.

Clark, Woodrow W., II. In press. *The Next Economics: Contemporary Economic Policy, Special Issue: Global Cases in Energy, Environment, and Climate Change: Some Challenges for the Field of Economics.* Fullerton, CA: International Western Economic Association.

Clark, Woodrow W., II, ed. 2010. *Sustainable Communities Design Handbook.* New York: Elsevier.

Clark, Woodrow W., II, and Ted Bradshaw. 2004. *Agile Energy Systems: Global Lessons from the California Energy Crisis.* Amsterdam: Elsevier.

Clark, Woodrow W., II, with RaeKwon Chong, dir. 1999. *Transfer of Environmentally Sound Technologies from Developed to Developing Countries.* Framework Convention for Climate Change, United Nations, New York.

Clark, Woodrow W., II, and Grant Cooke. 2010. The Third Industrial Revolution. In *Sustainable Communities Design Handbook,* edited by Woodrow W. Clark II, 9–22. New York: Elsevier.

Clark, Woodrow W., II, and Michael Fast. 2008. *Qualitative Economics: Toward a Science of Economics.* Chipping Norton, UK: Coxmoor Publishing.

Clark, Woodrow W., II, and William Isherwood. 2010. "Report on Energy Strategies for the Inner Mongolia Autonomous Region." *Utilities Policy* 18(1): 3–10. http://www.sciencedirect.com/science/article/pii/S0957178709000290.

Clark, Woodrow W., II, and Xing Li. 2003. "Social Capitalism: Transfer of Technology for Developing Nations." *International Journal of Technology Transfer* 3: 1–11.

Clark, Woodrow W., II, and Russell Vare. 2009. Introduction. In *Sustainable Communities,* pages 1–12. New York: Springer.

Clark, Woodrow W., II Author and Editor. Sustainable Communities Design Handbook. 2010. In *Sustainable Communities Design Handbook,* pages 1–12. Amsterdam: Elsevier.

"Collapse of Modern Economic Theory." 2009. *The Economist,* July 16.

Dodge, Darrel M. 2009. *Illustrated History of Wind Power: Early History through 1875.* Littleton, CO: TelosNet. http://telosnet.com/wind/index.html.

Dole, Malcolm. 2010. "Market Solutions for Climate Change Raise Basic Economic Questions." Unpublished paper, CARB, CAL EPA, Sacramento, CA.

Douglas, William. 2011. "Tax Breaks Still Intact for Big Oil." McClatchy Newspapers, May 18.

Dugar, Jatan, Phil Ting, and Johanna Partin. 2009. "Renewable Energy Practices in the City and County of San Francisco." In *Sustainable Communities,* edited by Woodrow W. Clark II, 95–108. New York: Springer.

Duggan, Jill. 2009. "The Truth About Cap-and-Trade in Europe." World Resources Institute, November 19. http://www.wri.org/stories/2009/11/truth-about-cap-and-trade-europe.

Duray, Dan. 2007. "Bush Endorses Climate Study." *Monterey County Herald,* February 3.

Energy Independence and Security Act of 2007. Public Law 110–140. http://frwebgate.access.gpo.gov/cgi-bin/getdoc.cgi?dbname=110_cong_public_laws&docid=f:publ140.110.

Energy Sector Investing. 2011. "China Oil Demand 9.2 Barrels Per Day," April 25. http://www.energysectorinvesting.com/2011/04/china-oil-demand-92m-barrels-per-day.html.

Environmental Defense Fund. 2010. "Prop. 23: Polluter's Dirty Energy Proposition for California," November 3. http://www.edf.org/page.cfm?tagID=58679.

Environmental and Energy Study Institute (EESI) and ChinaFAQs project of the World Resources Institute (WRI). 2011. "China's Energy and Climate Initiatives: Successes, Challenges, and Implications for US Policies." April 5 briefing SVC 203/202 Capitol Visitor Center. http://www.eesi.org/china%E2%80%99s-energy-and-climate-initiatives-successes-challenges-and-implications-us-policies-05-apr-2011.

Federal Ministry for the Environment, Nature Conservation and Nuclear Safety, Germany. 2000. "Act on Granting Priority to Renewable Energy Sources: Renewable Energy Sources Act," March. http://www.wind works.org/FeedLaws/Germany/GermanEEG2000.pdf.

Feldman, Stacy. 2010. "Algae Fuel Inches toward Price Parity with Oil." Reuters, November 22. http://www.reuters.com/article/idUS108599411820101122?pageNumber=title=Algae.

Food and Agriculture Organization of the United Nations. 2006. "Livestock's Long Shadow–Environmental Issues and Options." http://www.fao.org/ag/magazine/0612sp1.htm.

"France: Energy Profile." 2007. *EnerPub,* June 8. http://www.energypublisher.com/article.asp?id=9839.

Friedman, Thomas. 2007. "The Green-Collar Solution." *New York Times,* October 17.

Friedman, Thomas. 2008. *Hot, Flat and Crowded.* New York: Farrar, Straus and Giroux.

Funaki, Kentaro, and Lucas Adams. 2009. "Japanese Experience with Efforts at the Community Level towards a Sustainable Economy." In *Sustainable Communities,* edited by Woodrow W. Clark II, 243–62. New York: Springer.

Geothermal Energy Association. 2010. "Geothermal Energy International Market Update," May. http://www.geo-energy.org/pdf/reports/GEA_International_Market_Report_Final_May_2010.pdf.

Gipe, Paul. 2010–2011. "The Gipe Report: Feed-in-Tariff News." http://www.altaterra.net/members/blog_view.asp?id=511177.

Gipe, Paul. 2011. "Fukushima Nuclear Year-to-Year Reliability and German Wind," April 14. http://www.wind-works.org/FeedLaws/Japan/FukushimaNuclearYeartoYearReliabilityandGermanWind.html.

Goldman Sachs. 2009. *Environmental Progress Report.* http://www2.goldmansachs.com/services/advising/environmental-markets/documents-links/env-report-2009.pdf.

Greenpeace. 2011. "China Becomes World's Number 1 in Wind Installation," January 12. http://www.greenpeace.org/eastasia/press/release/china-world-leader-wind-energy.

Habermerier, Claus. 2009. "The Feed-in-Tariff System in Germany." Germany Trade and Invest, March. http://www.gtai.com.

Hamilton, Tyler. 2007. "Fresh Alarm over Global Warming." *The Toronto Star,* January 1. http://www.thestar.com/article/166819.

Hamilton, Tyler. 2010. "CO_2-Eating Algae Turns Cement Maker Green." *The Toronto Star,* March 18. http://www.thestar.com/business/article/781426—co2-eating-algae-turns-cement-maker-green.

Hansen, James. 2007. "Direct Testimony of James E. Hansen." State of Iowa, before the Iowa Utilities Board, November 5. http://www.columbia.edu/~jeh1/2007/IowaCoal_20071105.pdf.

Hansen, J. E., and S. Matsushima. 1967. "The Atmosphere and Surface Temperature of Venus: A Dust Insulation Model." *Astrophysical Journal* 150: 1139–57.

Hawkins, Paul, Amory Lovins, and L. Hunter Lovins. 1999. *Natural Capitalism: Creating the Next Industrial Revolution.* New York: Little, Brown and Company.

Helman, Christopher. 2010. "What the Top U.S. Companies Pay in Taxes." Forbes.com, April 1. http://www.forbes.com/2010/04/01/ge-exxon-walmart-business-washington-corporate-taxes.html.

Hoexter, Michael. 2008. "Cheap Energy Contract: A Critical Roadblock to Effective Energy Policy." *Green Thoughts,* January 14. http://greenthoughts.us/ree/cheapenergycontract/.

Hubbert, Marion King. 1956. "Nuclear Energy and the Fossil Fuels 'Drilling and Production Practice.'" Spring meeting of the Southern District, Division of Production. American Petroleum Institute, San Antonio, Texas. Shell Development Company, June 22–27. http://www.hubbertpeak.com/hubbert/1956/1956.pdf.

InterAcademy Panel on International Issues. 1994. "Statement on Population Growth." http://www.interacademies.net/3419.aspx?Redirect=3547.
Intergy Corporation. 2005. "Sustainability and Sustainable Communities." http://www.intergycorp.com/sustainable_services.html.
International Energy Agency. 2010. "Key World Energy Statistics." http://www.iea.org/textbase/nppdf/free/2010/key_stats_2010.pdf.
International Energy Agency Report to the United Nations. 2010. "Worldwide Engagement for Sustainable Energy Strategies," September. http://www.iea.org/publications/free_new_Desc.asp?PUBS_ID=2181.
Jha, Alok. 2008. "China's Clean Revolution." *Guardian* (London), August 1. http://www.guardian.co.uk/environment/2008/aug/01/renewableenergy.climatechange.
Jin, A. Jerry. 2010. "Transformational Relationship of Renewable Energies and the Smart Grid." In *Sustainable Communities Design Handbook: Greening Engineering, Architecture, and Technology*, edited by Woodrow W. Clark II, 217–32. New York: Elsevier.
Jno, Straun, T. Robertson, and John Markham. 2007. *The Regenerative Braking Story*. State College, PA: Venture Publications.
Kaplinsky, Raphael. 2006. "Revisiting the Revisited Terms of Trade: Will China Make a Difference?" *World Development* 34(6): 981–95.
Karl, Thomas R., Jerry M. Melillo, and Thomas C. Peterson, eds. 2009. *Global Climate Change Impacts in the United States.* U.S. Global Change Research Program. Cambridge: Cambridge University Press. www.globalchange.gov/usimpacts.
Kawasumi, Tetsuo. 1999. "Reconsidering John Manjiro." *The Manjiro Society*. http://www.manjiro.org/reconsider.html.
Kim, JeongIn. 2010. "Korean Economy: Past and the Future." Unpublished paper.
Kwan, Calvin Lee. 2009. "Rizhao: China's Green Beacon for Sustainable Chinese Cities." In *Sustainable Communities*, edited by Woodrow W. Clark II, 215–22. New York: Springer.
LaMonica, Martin. 2010. "Ernst & Young: China Clear Leader in Renewable Energy." *CNET News*, December 1. http:/www.news.cnet.com/8301–11128_3–20024232–54.html.
Lee Sang-Kyu, et al. 2010. "Industrial Strategies on the Convergence of IT and the Manufacturing Industry." KIET Occasional Paper No. 81, December.
Li, Minqi. 2010. "Peak Energy and the Limits to Economic Growth: China and the World." In *The Rise of China and the Capitalist World Order*, edited by Li Xing, 117–34. London: Ashgate.
Li, Xing, and Woodrow W. Clark. 2009. "Globalization and the Next Economy: A Theoretical and Critical Review." In *Globalization and Transnational Capitalism: Crises, Challenges and Alternatives*, edited by Li Xing, 83–107. Aalborg, Denmark: Aalborg University Press.
Li, Xing, and Woodrow W. Clark. 2010. "Energy Concern in China's Policy-Making Calculation: From Self-reliance, Market-dependence to Green Energy." Research

Center on Development and International Relations. DIR & Department of Culture and Global Studies Aalborg University. http://vbn.aau.dk/files/40057939/DIR_wp_143.pdf.

"Livability Rankings." 2010. *The Economist,* February 11. http://www.economist.com/blogs/gulliver/2010/02/liveability_rankings.

Lo, Vincent. 2011. "China's 12th Five-Year Plan." Speech given by Chairman Shui On Land and President of the Yangtze Council. Asian Society Meeting, Los Angeles, California, April 25.

Lund, Henrik, and Poul Alberg Ostergaard. 2009. "Sustainable Towns: The Case of Frederikshavn—100% Renewable Energy." In *Sustainable Communities*, edited by Woodrow W. Clark II, 155–68. New York: Springer.

"Major Power Outage Hits New York, Other Large Cities." 2003. *CNN*, August 14. http://www.cnn.com/2003/US/08/14/power.outage/.

National Institute of Standards and Technologies. 2005. "Drivers of Innovation in the Fuel Cell Industry." NISTIR 7161. http://www.atp.nist.gov/eao/ir-7161/chapt3.htm.

National Labor Report. 2010. *Annual Report: Fiscal Year 2009*. http://www.nlrb.gov/sites/default/files/documents/119/nlrb2009.pdf.

Nelson, Richard. 2011. "Concentrated Solar Electric Energy." *Renewable Power News,* April 7. www.renewablepowernews.com/archives/1220.

"New York Orders Flywheel Energy Storage." 2009. *Railway Gazette*. http://www.railwaygazette.com/nc/news/single-view/view/new-york-orders-fly-wheel-energy-storage.html.

Ostergaard, Poul Alberg, and Henrik Lund. 2010. "Climate Change Mitigation from the Bottom-up Community Approach." In *Sustainable Communities Design Handbook*, edited by Woodrow W. Clark II, 247–66. New York: Elsevier.

O'Toole, Sarah. 2010. "World's Biggest Offshore Wind farm Opens Today." *Global Energy Magazine*, September 23. www.globalenergymagazine.com/2010/09/worlds-biggest-offshore-wind-farm-opens-today/.

Pacific Northwest National Laboratory. 2003. "GridWiseTM: The Benefits of a Transformed Energy System," September. http://arxiv.org/ftp/nlin/papers/0409/0409035.pdf.

Padgett, Tim, and Stephen Kuffner. 2011. "Chevron v. Ecuador: Will the Plaintiffs Get Paid?" *Time*, March. http://www.time.com/time/world/article/0,8599,2053075,00.html.

"Parry People Movers for Stourbridge Branch Line." 2010. *London Midland*, March 1. http://web.archive.org/web/20080517110918/http://www.londonmidland.com/index.php/news/news_items/view/23.

Pew Charitable Trusts. 2010. "Who's Winning the Clean Energy Race? Growth, Competition and Opportunity in the World's Largest Economies." *G-20 Clean Energy Fact Book*. http://www.pewtrusts.org/uploadedFiles/wwwpewtrustsorg/Reports/Global_warming/G-20%20Report.pdf.

Pew Chartable Trusts. 2011. "Global Clean Energy Investment Reached Record $243 Billion in 2010," March 29. http://www.pewenvironment.org/news-room/

press-releases/global-clean-energy-investment-reached-record-243-billion-in-2010–329326.

Pinderhughes, Raquel. 2006. "Green Collar Jobs: Work Force Opportunities in the Growing Green Economy." Urban Habitat / Race Poverty and the Environment (Summer). http://urbanhabitat.org/files/Pinderhughes.Green.Jobs.pdf.

Pollin, Robert, Heidi Garrett-Peltier, James Heintz, and Helen Scharber. 2008. "Green Recovery: A Program to Create Good Jobs and Start Building a Low-Carbon Economy." Center for American Progress and Political Economy Research Institute (September). http://www.peri.umass.edu/fileadmin/pdf/other_publication_types/peri_report.pdf.

PVResources. 2011. "A Walk Through Time—Discoveries of Basic Phenomena and Properties of PV Materials," January. www.pvresources.com/en/history.php.

"Q&A: The U.S. and Climate Change." 2002. *BBC News*, February 14. http://news.bbc.co.uk/2/hi/americas/1820523.stm.

Renewable Energy Policy Network for the 21st Century (REN21). 2009. "Renewables Global Status Report: 2009 Update." http://www.ren21.net/Portals/97/documents/GSR/RE_GSR_2009_Update.pdf.

"Report: Clean Energy Trends 2011." 2011. CleanEdge.com. http://www.cleanedge.com/reports/reports-trends2011.php.

Rifkin, Jeremy. 2004. *The European Dream*. New York: Tarcher/Penguin.

Rifkin, Jeremy. 2007. "The EU Leads the World in Renewable Energy." Speech in Rome to the mayor and city council, March 27.

Robertson, Campbell, and Leslie Kaufman. 2010. "Size of Spill in Gulf of Mexico Is Larger Than Thought." *New York Times* (New York edition), April 29, A14.

Roth, Matthew. 2010 "New Report Impugns Texas Oil Companies Funding California Prop 23." *San Francisco Chronicle*, August 10.

Samson, R., and S. Bailey Stamler. 2009. "Going Green for Less: Cost-Effective Alternative Energy Sources." *C.D. Howe Institute Commentary* 282 (February).

Scott, Mark. 2009. "How Italy Beat the World to a Smarter Grid." *Bloomberg Businessweek*, November 16. http://www.businessweek.com/globalbiz/content/nov2009/gb20091116_319929.htm.

Sears, Richard. 2010. "Planning for the End of Oil." Presentation at the Technology Entertainment and Design Conference, February, California.

Shabecoff, Philip. 1988. "Global Warming Has Begun, Expert Tells Senate." *New York Times*, June 24. http://www.nytimes.com/1988/06/24/us/global-warming-has-begun-expert-tells-senate.html.

Shoumatoff, Alex. 2008. "The Arctic Oil Rush." *Vanity Fair*, May. http://www.vanityfair.com/politics/features/2008/05/arctic_oil200805.

South Coast Air Quality Management District. 2010. "Final Environmental

Spencer, Nicole, and Siobhan Sheils. 2009. "Chile Update: Energy Policy." Americas Society, September 9. www.as-coa.org/articles/1879/Chile_Update:_Energy_Policy/sitemap.php.

Texas Transportation Institute. 2005. "Urban Mobility Report." http://mobility.tamu.edu/ums/.

Thøgersen, Stig, and Clemens S. Østergaard. 2010. "Chinese Globalization: State Strategies and Their Social Anchoring." In *The Rise of China and the Capitalist World Order*, edited by Li Xing, 161–86. London: Ashgate.

Tickell, Josh, dir. 2009. *Fuel*. Documentary film. Los Angeles, CA: La Cinema Libra.

Tokarz, Frank. 1996–1998. "Regenerative Braking Intellectual Property," Energy directorate, technology transfer. Lawrence Livermore National Laboratory. Internal memos and e-mails, University of California.

Tollefson, Jeff. 2011. "America's Top Climate Cop," May 19. http://www.nature.com/news/2011/110518/pdf/473268a.pdf.

"UK Scientists' IPCC Reaction." 2007. *BBC News*, February 2. http://news.bbc.co.uk/2/hi/science/nature/6324093.stm.

UN Department of Economic and Social Affairs. 2011. "World Population Prospects: The 2010 Revision." http://esa.un.org/unpd/wpp/index.htm.

UN Environment Programme. 2008. "Green Jobs: Towards Decent Work in a Sustainable, Low-Carbon World," September. http://www.unep.org/labour_environment/PDFs/Greenjobs/UNEP-Green-Jobs-Report.pdf.

UN Environment Programme. 2009. "Global Trends in Sustainable Energy Investment 2009." http://www.unep.org/pdf/Global_trends_report_2009.pdf.

UN Intergovernmental Panel on Climate Change. 1990. *First Assessment Report*. Summary. http://www.ipcc.ch/publications_and_data/publications_and_data_reports.shtml.

UN Intergovernmental Panel on Climate Change. 1995. *Second Assessment Report*. Summary. http://www.ipcc.ch/publications_and_data/publications_and_data_reports.shtml.

UN Intergovernmental Panel on Climate Change. 2001. *Third Assessment Report*. Summary. http://www.ipcc.ch/publications_and_data/publications_and_data_reports.shtml.

UN Intergovernmental Panel on Climate Change. 2007. *Fourth Assessment Report*. Summary. http://www.ipcc.ch/publications_and_data/publications_and_data_reports.shtml.

Union for the Conservation of Nature. 2009. "Update On the IUCN Red List of Threatened Species." News release, November 3. http://www.iucn.org/about/work/programmes/species/red_list/?4143/Extinction-crisis-continues-apace.

Union of Concerned Scientists. 2007. "Fuel Economy Basics," November 26. http://www.ucsusa.org/clean_vehicles/solutions/cleaner_cars_pickups_and_suvs/fuel-economy-basics.html#History_of_US_Fuel_Economy_Standards.

Union of Concerned Scientists. 2010. "How Biomass Energy Works," October. http://www.ucsusa.org/clean_energy/technology_and_impacts/energy_technologies/how-biomass-energy-works.html.

United Nations. 2010. "Revision of World Population Prospects." www.un.org/esa/population/unpop.htm.

U.S.–Canada Power System Outage Task Force. 2004. "Final Report on the August 14, 2003 Blackout in the United States and Canada: Causes and Recommendations," April. https://reports.energy.gov/BlackoutFinal-Web.pdf.

U.S. Department of Commerce. 2011. "Greenhouse Gases." The National Environmental Satellite, Data, and Information Service. http://www.ncdc.noaa.gov/oa/climate/gases.html#introduction.

U.S. Department of Energy. "Annual Global Reports." http://management.energy.gov/documents/annual_reports.htm.

U.S. Department of Energy. 2001. "Fuels Reserve-Production Ratio Report and Charts." Energy Information Administration.

U.S. Department of Energy, Energy Information Administration. 2011. "Annual Energy Outlook 2011," December 16. http://www.eia.gov/forecasts/aeo/MT_energydemand.cfm.

U.S. Department of the Interior, National Park Service. 2004. *Climate Change in the National Parks.*

U.S. Department of the Interior, National Park Service. 2011. "Secure Water Act Report." http://www.usbr.gov/climate.

U.S. Geological Survey. 2003. "USGS World Petroleum Assessment 2000." USGS Fact Sheet FS–062–03, June. http://pubs.usgs.gov/fs/fs-062–03/FS-062–03.pdf.

U.S. Information Technology and Innovation Foundation. 2009. "Rising Tigers, Sleeping Giant: Asian Nations Set to Dominate the Clean Energy Race by Out-Investing the United States," November. http://thebreakthrough.org/blog/Rising_Tigers_Summary.pdf.

Venture Business Research Limited. 2010. "CleanTech and Renewable Energy Review 2Q10," July 12. http://www.smartgridnews.com/artman/uploads/1/clean_tech_Q2.pdf.

Walton, Marsha. 2008. "Algae: The Ultimate in Renewable Energy." *CNNTech,* April 1. http://articles.cnn.com/2008–04–01/tech/algae.oil_1_algae-research-fossil-fuels-nrel?_s=PM:TECH.

Wesoff, Eric. 2010. "Milestone 10 GW of Solar Panels in 2010." Greentechmedia.com, October 6. www.greentechmedia.com/articles/read/10-gigawatts/.

While, Franny. 2011. "Study: Algae Could Replace 17% of U.S. Oil Imports." *Renewable Energy World*, April 20. http://www.renewableenergyworld.com/rea/news/article/2011/04/study-algae-could-replace-17-of-u-s-oil-imports.

"World: Now No.2, Could China Become No.1?" 2011. *Time*, February 28: 17. http://www.time.com/time/magazine/article/0,9171,2050025–2,00.html.

World Wind Energy Association. 2011. "Wind Energy Report 2010 Executive Summary," April. www.wwindea.org/home/images/stories/pdfs/worldwindenergyreport2010_s.pdf.

Yang, Robert. 2010. "Business Opportunities: The Key to Any Nation's Green Energy Industry Strategy." Smart Green Cities Summit, Stanford Program on Regions of Innovation and Entrepreneurship (SPRIE). Keynote presentation, Panel 5, May 10, Stanford University, California.

Zhao, Zhen Yu, Ji Hu, and Jian Zuo. 2009. "Performance of Wind Power Industry Development in China: A Diamond Model Study." *Renewable Energy Journal* 34: 2883–91.

Index

Abe, Shinzo, 44
ABI Research, 123
Active-matrix organic light-emitting diode, 48
Afghanistan war, 21, 150
Africa, renewable energy in, 114–15
Age of Enlightenment, 4, 172
Agile Energy Systems, Global Lessons from the California Energy Crisis (Clark & Bradshaw), 15
Agile sustainable communities, 59–60, 83–84, 94–95
Algae, as a biofuel, 143–45
Algae Fuels S.A. (Chile), 85
Algal Biomass Organization, 144
Algal Collection, 143
al-Qaeda, 150
AMC Amitron (1967) regenerative braking car, 140
American Lung Association, 78–79
American Wind Energy Association, 102
Animal feed lots: biomass energy production and, 109–10. *See also* Livestock
An Inconvenient Truth (Gore), 13, 23
"Annual Energy Outlook" (DOE, Energy Information Administration, 2011), 97–98
Anthropogenic warming factors, 33, 35
Apple Computer, 129
Arab oil embargoes (early 1970s), 151, 169
ARCO Solar, 104
Arctic Climate Impact Assessment, 25
Arctic Council and the International Arctic Science Committee, 23
Arctic Monitoring and Assessment Program, 29
Arctic Rim countries, 29
ASM, 60, 61
Atmosphere changes, 33

Bacterial fuel cells, 18–19
Becquerel, Alexandre Edmond, 104
Beijing Olympics (2008), 63–64, 66
Bell, Alexander Graham, 4
Berkeley First (California), 106
BESCO, 54
Biofuels, 143–45
Biofuels Feedstock Development Program, 145–46
Biogas, 109
Biogenerator, 18
Biomass, 17, 82, 109–10, 138
Blackouts: in France and Italy, 119–20; in U.S. and Canada (2003), 119, 120
Black Ships, 42
Bloomberg Businessweek, 123
Bloom Energy fuel cell, 136
Boulton, Matthew, 4
BP. *See* British Petroleum
Brazil: investment in renewable energy (2009), 114; sugar cane ethanol in, 85, 169; Tupi Field, 9–10
British Broadcasting Corporation, 36
British Petroleum (BP), 10; Gulf of Mexico oil spill (2010), 11–12, 18–19; investment in bacterial/microbial fuel cells, 18
British Petroleum America, 151
Brown, Jerry, 13, 172
Brundtland Commission (UN), 80
Brush, Charles F., 100
Bush, George H.W., 170
Bush, George W., 24, 36, 170, 171

California: agile sustainable communities and, 84; energy efficient programs in, 21, 164, 172; G3IR and, 138; Proposition, 23, 13; Renewable Auction Mechanism (RAM) (Public Utility Commission) FiT style program, 157, 166; solar power in, 105–6
California Air Resources Board: cap-and-trade bill, 160; Toyota Prius rating, 94
Callahan, David, *Fortunes of Change: The Rise of the Liberal Rich and the Remaking of America*, 3
Capacity management, 132–34
Cap and trade, 157–61

Carbon credit trading, 58, 160–61. *See also* Cap and trade; Carbon tax
Carbon dioxide, 26–28, 35–36, 58, 144–45. *See also* Greenhouse gases
Carbonless economy, moving toward in U.S., 12–16, 167–78
Carbon tax, 159–60
Cars, 26–27; in China, 24. *See also* Hybrid technology; Hydrogen fuel cells
Carter, Jimmy, 169–70
CDM. *See* Clean Development Mechanism, 58
Cell phones, solar systems and, 114–15
Central-grid power plants (2IR), 19–20, 117–20, 132–33
Centralized economic policy, 65. *See also* China
Certification, 51; Leadership in Energy and Environmental Design (LEED), 82, 90, 91, 92
"Cheap Energy Contract: A Critical Roadblock to Effective Energy Policy in the U.S." (Hoexter), 150, 151, 156–57
Chemical combustion, 97
Chevron, 151, 167
Chicago School of supply-side economics, 55
Chile: Algae Fuels S.A., 85; renewable energy in, 85; wind farms in, 102–3
China, 85, 177; 12th 5-year plan, 63, 65–66, 153; 2008 Beijing Olympics, 63–64; cars in, 24; as emerging world leader in sustainability, 66–67; energy needs, 67–68; Feed-in-Tariff in, 71–72; government role in G3IR policies, 152–54; job generation and G3IR, 161–63; maglev in, 66, 141; oil consumption in, 10–11; as part of the green technology tigers, 72–75; pollution in, 20, 64; renewable energy investment in, 165–66; socialist capitalist economic model, 65; solar power in, 105, 106; Solar Valley City, 68–69; wind power in, 70–72; wind turbine manufacturing in, 14; workforce training for green technology, 64–65. *See also* Social capitalism
Chinese Renewable Energy Industries Association, 70
Cisco, 127
Cities, population in, 77
City Life (Italy), 61
Clark, Woodrow W., II and Michael Fast, *Qualitative Economics: Toward a Science of Economics*, 2–3
Clark, Woodrow W., II and Ted Bradshaw, *Agile Energy Systems, Global Lessons from the California Energy Crisis*, 15
Clean Air Act, 78
Clean Development Mechanism (CDM), 58
Clean Edge, 107
"Clean energy," 13
Clean rich, 3
Cleantech investment by country, 50
Climate change, 4–5, 7–8, 23–39, 174, 178; conclusions from the *Global Climate Change Impacts in the United States*, 37–39; defining, 25–29; findings of the *Fourth Assessment Report* (UN IPCC), 32–35; James Hansen's research on, 29–31; populations increases and, 39–40; United Nations Framework Convention on Climate Change (UN FCCC), 31–32; United Nations Intergovernmental Panel on Climate Change (UN IPCC), 31–32. *See also* Global warming
Climate Group, 12
Clinton, Bill, 35–36, 170–71
Coal, 13, 20, 26, 30, 68, 97–98, 168
Coal-burning steam engine, 4
Code division multiple access, 47
Cold War, end of, 53, 149
"Collapse of Modern Economic Theory," *The Economist*, 50, 51, 55, 149
Commercial technologies and G3IR, 129–48; biofuels, 143–45; commercializing emerging technologies, 146–48; energy conservation and efficiency technologies and, 130–34; flywheel energy storage (FES), 141–43; fuel cell storage, 134–39; fuel from plants, 145–46; hydrogen fuel cells, 136–39; magnetic levitation (magelv) high speed rail, 140–41; regeneration braking, 139–40. *See also* Renewable energy; Smart green grids
Communist Party of the National Government of China, 65
Communities. *See* Sustainable communities
Concentrated solar power (CSP) technologies, 106
ConocoPhillips, 151
Conservation, 51, 130-34
"Cool Earth 50" initiative, 44
Corn-produced energy, 146
Council of the European Union, emissions trading program, 58
Credit derivative meltdown, 53
Crop challenges, 38
CSP technologies. *See* Concentrated solar power (CSP) technologies
"Cyber Korea 21" (South Korea), 47

Data response, smart green grids and, 126–28
Definitions, 13–14
Deforestation, 28
Dell, Michael, 3
Denmark: 100 percent renewable power (by 2050), 56; energy independent cities in (by 2015), 8; Frederikshavn (city with renewable energy), 85; local onsite power in, 82; wind power manufacturing in, 56–57, 65, 154
Department of Commerce, 8
Department of Energy, 8–9, 136; Energy Information Administration "Annual Energy Outlook" (2011), 97–98; *GridWise®: The Benefits of a Transformed Energy System* (Pacific Northwest National Laboratory), 121–22; Information Administration, 10
Digital/information technology for electricity networks, 120. *See also* Smart green grids
Dimmable ballasts, 132, 164
Dirty rich, 3
Distributed control, smart green grids and, 127
Distributed energy generation, 59–62
Drivers for transformation, 2

Earthquake and tsunami in Japan (March 11, 2011), 11, 45–46
Economic crisis (2008), 8, 12, 91, 149, 161, 167
Economics: of 2IR, 149–51. *See also* Free market capitalism; Neoclassical economics; Social capitalism
Economist, The, "Collapse of Modern Economic Theory," 50, 51, 55, 149
Edison, Thomas, 19, 118–19
Efficiency technologies, 130–34, 176–77
EGP. *See* European Green Party
Electric automobiles, 139
Electric cars, 84
Electricite de France, 18
Electricity: delivery into homes, 118; distribution of, 118; forms of production, 118; historical background, 117–20; transmission of, 118, 120; UN estimates of world's population without, 114; use of in U.S., 97–98
Electric railways, regenerative braking and, 139–40
Electrolysis of water, 138–39
Electromagnetic radiation, 29
Electromechanical generators, 97
Elemental hydrogen, 138
Emissions trading programs, 58, 157–61. *See also* Cap and trade; Carbon tax
Emission Trading System (EU), 159
Endangered species, 5
ENEL (Italy), 123, 127
Energy: generated by fossil fuels in the U.S., 97–98; real cost of fossil fuels, 150–51. *See also* Green Third Industrial Revolution; Second Industrial Revolution
"Energy Concerns in China's Policy-making Calculations: From Self-Reliance, Market-Dependence to Green Energy" (Li & Clark), 153
Energy conservation, 130–34
Energy efficiency, 51
Energy management system, 133
Energy Policy and Conservation Act (1975), 169–70
Energy Renewable Sources Act (Germany, formerly the Act of Granting Priority to Renewable Energy Sources), 56, 155–56
Energy security, 154
Energy Star, 82
England, wind farms in, 102
Environmental Protection Agency, 36, 94, 137
Ernst & Young, 165–66
Ethanol, 169
EU. *See* European Union
Europe, Green Third Industrial Revolution (G3IR) in, 1, 7, 12, 53–62, 154–57
European Commission, 156
European Dream, The (Rifkin), 1, 7, 169
European Green Party (EGP), 54
European Union (EU), 54–59; Emission Trading System, 159; policies, 57–59; smart meters and, 122, 123–24
ExxonMobil, 130, 150–51, 167

Faraday, Michael, 97
Federal Energy Regulatory Commission (FERC), 119, 120
Federal stimulus funds, 14
Feed-in-Tariff (FiT): in California, 157; in China, 71–72; in Germany, 49, 56, 58–59, 105, 155–57, 164; in Japan, 44; worldwide, 156
FERC. *See* Federal Energy Regulatory Commission
FES. *See* Flywheel energy storage
Fifth Assessment Report (UN IPCC), 35
First Assessment Report (UN IPCC), 31–32
First Industrial Revolution (1IR), 3–4, 25–26, 149, 172

FiT. *See* Feed-in-Tariff
Fluorescent tubes, 131–32
Flywheel energy storage (FES), 141–43
Forbes, report to the Security Exchange Commission, 150–51
Ford, Gerals, 169
Fortunes of Change: The Rise of the Liberal Rich and the Remaking of America (Callahan), 3
Fossil-fuel fortunes, 3
Fossil fuels, 1;use in U.S., 97–98. *See also* Second Industrial Revolution (2IR)
Fourth Assessment Report (UN IPCC), 32–35
France, nuclear power in, 45, 169
Franklin, Ben, 26
Free market capitalism (economics of 2IR), 53, 149–51. *See also* Neoclassical economics
Friedman, Thomas, 168
Fuel cell storage, 18–19, 51, 134–39; defined, 134–35; fuel types for, 136; hydrogen fuel cells, 136–39; invention of, 135; workings of, 135–36
Fuel mileage standards, 170
Future Melbourne, 87–88

G3IR. *See* Green Third Industrial Revolution
Garbage, 109. *See also* Biomass
General Electric, 167
Geothermal energy, 15, 17–18, 108–9; converting into energy, 108; historical background, 108; in the U.S., 109; worldwide, 108
German Transrapid International, 141
Germany, 54; Feed-in-Tariff (FiT) and, 49, 56, 58–59, 155–57, 164; solar power in, 105
Geysers in Northern California, 108
GHGs. *See* Greenhouse gases
Global Climate Change Impacts in the United States, 24, 37–39
Global economy, 177
Global warming, 4–5, 7–8, 174, 178. *See also* Climate change
Goldman Sachs, 164–65
Google, 164–65
Gore, Al, 36; *An Inconvenient Truth*, 13, 23
Grand Coulee Dam, 110
Great Recession (2008), 8, 12, 91, 149, 161, 167
Green, defining, 13–14
Green Building Council, Leadership in Energy and Environmental Design (LEED), 82, 90–92
"Green Collar Jobs: Work Force Opportunities in the Growing Green Economy" (Pinderhughes), 163
Green collar jobs, 163–64. *See also* Green jobs
Green Growth Task Force (South Korea), 46
Greenhouse gases (GHGs), 16, 24, 27–39, 58, 98. *See also* Carbon dioxide; Climate change; Emissions trading programs; Methane; Nitrous oxide
Green Jobs: Towards Decent Work in a Sustainable Low-Carbon World (UN Environment Programme), 163
Green jobs, 2, 3, 56, 161–64. *See also* Green collar jobs
"Green Recovery: A Program to Create Good Jobs and Start Building a Low-Carbon Economy" (Pollin et al), 163–64
Green technology tigers (China, Japan, South Korea), 72–75, 105
Green Third Industrial Revolution (G3IR): the Asian approach and U.S. reaction, 48–52; background of, 1–5, 7; biomass: recycled and reusable generation, 109–10; China and, 63–75; drivers for transformation to, 2; Europe and, 53–62, 154–57; five key elements of, 51–52; geothermal energy, 108–9; historical precedence for shift from 2IR to (1970s Arab oil embargo), 151; Japan and, 41–46; moving toward in U.S., 7–22, 167–78; in poor and rural communities, 114–15; to protect the environment and change local communities, 113–16; solar energy, 104–8; South Korea and, 46–48; water as energy, 110–12; wind power generation, 99–104. *See also* Commercial technologies and G3IR; Feed-in-Tariff (FiT); Renewable energy; Smart green grids; Social capitalism; Sustainable communities
Grid parity, 56
GridWise®: The Benefits of a Transformed Energy System (Pacific Northwest Laboratory for the DOE), 121–22
Ground Zero (New York City), 171
Grove, William, 135

Halocarbons, 35
Hansen, James, climate change studies, 29–31
Hewlett and Packard, 171
Hiroshima, nuclear bombing of, 42
Hoexter, Michael, "Cheap Energy Contract: A Critical Roadblock to Effective Energy Policy in the U.S.," 150, 151, 156–57
Holland, 54
Home/family, green revolution's/sustainability starting place, 61–62, 80, 84–86

Honda FCX Clarity, 137–38
Hoover Dam, 110
Hotels/resorts, sustainability and, 90–92
Hubbert, M. King, 9
Hurricane Katrina, 174
Hurricanes, increase in, 34
Husk Power System (India), 114
Hybrid technology, 8–9, 84–85, 137, 139
Hydroelectric power, 110–12
Hydrofluorocarbons, 35
Hydrogen Association of America, 136
Hydrogen cars, 136–38
Hydrogen fuel cells, 18, 94, 136–39

Ice sheets, diminishing, 34
IGA. *See* International Geothermal Association
INAX, 45
Incandescent bulbs, 131
Incentives. *See* Tax breaks; Subsidies; Emissions trading programs
Incident days, 133–34
Incredible Hulk roller coaster, 141
India: Husk Power Systems (rice husks), 114; investment in renewable energy (2009), 114; oil consumption in, 11
Information Technology and Innovation Foundation, *Rising Tigers, Sleeping Giants*, 72
Information technology bubble, 171
Infrastructure elements for sustainability, 81
Intermittent technologies, 19–21. *See also* Storage technologies
Internal combustion engines, 1, 4, 137
International Energy Agency, 156
International Geothermal Association (IGA), 108
International Union for Conservation of Nature (IUCN), 5
Internet: distributed control and, 127; smart operating systems and, 134
Internet bubble, 171
Internet-connected dimmable ballast, 132
Invisible hand, 149
IOUs, 127
Iraq war, 21, 150
Irreversible environmental degradation, 173
Islamic fundamentalism, 53
Italy: onsite power systems, 60–61; smart meters in, 123
IUCN. *See* International Union for Conservation of Nature

Japan: earthquake and tsunami in (March 11, 2011), 11, 45–46; G3IR in, 1–2, 7, 12, 41–46, 79, 85–86; nuclear bombing of, 42; solar power in, 105
Japanese-Russian War (1905), 42
Jobs, Steve, 129
Joint Implementation, 58

Kan, Naota, 45
Kinetic energy, 97
Klug, Astrid, 164
Knowledge-based jobs, 163
Krieger, Louis Antoine, 139
Kyoto Protocol, 24, 35–37, 44, 58

LACCD. *See* Los Angeles Community College District
Langer, Charles, 135
Lawrence Berkeley laboratory, 145
Lawrence Livermore laboratory, 145
Leadership in Energy and Environmental Design (LEED) certification, 82, 90–92
LEDs, 134, 164
LEED. *See* Leadership in Energy and Environmental Design
Lemnis Lighting, 134
Liberalization, 59
"Lifeworld," 50
Lighting, 131–32
Liquefied natural gas, 130
Livestock, greenhouse gases and, 28–29
Livestock challenges, 38
Livestock Information and Policy Branch (UN Food and Agriculture Organization), 28
Livestock's Long Shadow—Environmental Issues and Options (UN), 28–29
Li, Xing & Woodrow W. Clark, "Energy Concerns in China's Policy-making Calculations: From Self-Reliance, Market-Dependence to Green Energy," 153
Local power generation, 124–26. *See also* Commercial technologies and G3IR; Onsite power systems; Smart green grids; Sustainable communities
Lohner-Porsche Carriage, 84
Los Alamos National Laboratory, 136
Los Angeles car culture, 78–79
Los Angeles Community College District (LACCD), 89–90

Maglev. *See* Magnetic levitation (maglev) high-speed rail lines
Magnetic levitation (maglev) high-speed rail lines, 66–67, 140–41

Manjiro, Nakahama, 42
Manure, 109–10
Marine water currents, 111
Market-capitalism system, 153–54. *See also* Social capitalism
Mechanical engineering, 1
Megacities, 77
Megatrend (G3IR), 168
Melbourne, Australia, 86–89
Melbourne Principles, 94
Melting ice caps, 5
Metabolic engineering, 143
Methane, 27–28, 35
Methyl Tertiary Butyl Ether (MTBE), 79
Meucci, Antonio, 4
Microbial fuel cells, 18–19
Middle class, in developing nations, 172–73
Middle East, turmoil and revolutions in, 8, 150, 167, 174
Mobile dynamic random-access memory (DRAM), 48
Modern economic theory, 50, 55
Mond, Ludwig, 135
Monopolies, 19–20
Montreal Protocol (1987), 36
MTBE. *See* Methyl Tertiary Butyl Ether

Nagasaki, nuclear bombing of, 42
NAP. *See* National allocation plans, 58
National Aeronautics and Space Administration (NASA) Goddard Institute for Space Studies, 29
National allocation plans (NAP), 58
National policies on G3IR, 152, 160, 175
National Renewable Energy Laboratory, 144
Natural gas, 9, 13, 97, 98
Neoclassical economics, 1, 50, 151, 156–57
Net metering, 124–25
New York, energy efficient programs in, 21, 172
Next economics. *See* Social capitalism
Nichols, Mary, 160
Nitrous oxide, 27–29, 35
Nixon, Richard, 169
North Sea oil deposits, 54
Norway, 23
Nuclear fission, 97
Nuclear power plants, 45; in France, 169; Japan's earthquake and tsunami disabling of (March, 2011), 11, 45–46; in Russia, 45

Oak Ridge National Laboratory, 145
Obama, Barak, 25, 36–37; Solar America Initiative, 107
Ocean and tidal wave powers, 18, 111–12. *See also* Water energy
Ocean oil spill, by British Petroleum (2010), 11–12
Ocean warming, 34
Oil: declining supplies of, 9–12; defined, 9. *See also* Climate change; Second Industrial Revolution (2IR)
Oil spill, by British Petroleum in Gulf of Mexico (2010), 11–12, 18–19
Oil wars, 11
1IR. *See* First Industrial Revolution
Onsite power systems, 60, 81–82, 130–31. *See also* Commercial technologies and G3IR; Smart green grids; Sustainable communities
Organization for Economic Cooperation and Development (OECD) countries, 48
Our Common Future (UN), 80
Overload, 133
Ozone laundry conversions, 93
Ozone layer, 36

Peak-load management, 133–34
Perennial grasses, 146
Perfluorocarbons, 35
Permafrost, diminishing, 29, 34
Perry, Matthew, 42
Photosynthesis, 110
Photovoltaic (PV) systems, 12, 67, 69, 104–5. *See also* Solar energy
Pinderhughes, Raquel, "Green Collar Jobs: Work Force Opportunities in the Growing Green Economy," 163
Planet warming, 34–35
Plants, fuel from, 143–46
Platts, 10
Political Economy Research Institute, 163
Pollin, Robert et al., "Green Recovery: A Program to Create Good Jobs and Start Building a Low-Carbon Economy," 163–64
Pollution: in China, 20, 64; from fossil fuels (2IR), 13, 130; in Los Angeles, 78–79. *See also* Climate change; Greenhouse gases (GHGs); Second Industrial Revolution
Population: in cities, 77; growing, 39–40, 172
Positive forcing, 35
Private investment, in clean/green technologies, 164–66
Private partnerships, 147
Privatization, 59
Procurement, 147
Public partnerships, 147
PV systems. *See* Photovoltaic (PV) systems

Qualitative Economics: Toward a Science of Economics (Clark & Fast), 2–3
Qualitative economics, 50

Radiative energy transfer, 29
Radiative forcing, 35
Railways: regenerative braking and, 139–40. *See also* Magnetic levitation (maglev) high-speed rail lines
Rationing, 133
Reagan, Ronald, 53, 149, 170
Recession of 2008. *See* Economic crisis (2008)
Regeneration braking, 139–40
Regulations, 149: of central electric grids, 119, 120
Renewable Auction Mechanism (RAM) (California's Public Utility Commission), 157, 166
Renewable energy, 1–3, 16–19, 51, 173–74; biomass, 109–10; geothermal, 108–9; investment in worldwide, 113–14; to protect the environment and change local communities, 113–16; solar energy, 104–8; water as energy, 110–12; wind power generation, 99–104. *See also* Headings under specific types
Renewable Energy Laws, 54
Renewable Global Status Report: 2009 Update, FiT worldwide, 156
Resorts, sustainability and, 90–92
Responsive dimmable ballasts, 164
Retribution, of George W. Bush after 9/11, 171
Retrofitting chemical buildings (California), 164
Rice husk power (India), 114
Rifkin, Jeremy, *European Dream, The*, 1, 7, 169
Rising Tigers, Sleeping Giants (Information Technology and Innovation Foundation), 72
River energy, 112
Roberts Enterprise Development Fund, 163
Rocky Mountain Climate Organization, 25
Roscoe Wind Farm (Texas), 102
Russia, nuclear power plants in, 45

Salt layers, 9
Samuri, 42
Sandia national laboratory, 145
San Francisco, 86, 92–93
Santa Monica, 86
Scandinavians, 154–55
Schools, sustainability and, 89–90
SeaGen, 18
Sea levels, rising, 34, 38
Sears, Richard, 10
Second Assessment Report (UN IPCC), 32
Second Industrial Revolution (2IR), 1, 2, 4, 7, 11, 20, 26, 30, 42–43, 77–79, 162, 167, 172; central power grids and, 117–20; economics of, 149–51
Security Exchange Commission, *Forbes* report to, 150–51
September 11 2001 terrorist attacks, 1, 49, 150; post 9/11 in U.S., 170–73
Shale oil, 98
Shanghai Maglev train, 66
Sharp Corporation, 104
Shell Oil, 9, 10, 151
Shelter cash, 151
Shogun (Japanese leaders), 42
Silicon, 17
Silicon cells, 104
Silicon crystal, 105
Silicon Valley (California), 150
Smart communities. *See* Sustainable communities
Smart green grids, 1, 2, 51, 120–28; background of, 121–22; data response and power transmission lines, 126–28; defining, 121, 122; distributed control concept (Internet) and, 127; efficient power transmission and distribution of, 120–26; *GridWise®: The Benefits of a Transformed Energy System* (Pacific Northwest National Laboratory study for the DOE), 121–22; local power generation, 124–26; net metering, 124–25; obstacles to, 126; smart meters, 122–24. *See also* Commercial technologies and G3IR; Onsite power systems; Sustainable communities
Smart meters, 122–24
Smith, Adam, 50
Smith, Willoughby, 104
Social capitalism, 50, 63, 65, 113, 177; cap and trade, 157–61; carbon taxes, 159–60; in China, 152–54; defined, 152; in Europe, 154–57; FiT in California, 157; Germany's FiT strategy, 155–57; green jobs and G3IR, 161–64; nations following the FiT model, 156; private investment and clean/green technology, 164–66. *See also* Commercial technologies and G3IR; Feed-in-Tariff (FiT); Green Third Industrial Revolution (G3IR); Renewable energy; Smart green grids; Sustainable communities
Social media revolution, 167
Solar America Initiative (Obama), 107
Solar cells (silicon), 17

Solar energy, 15, 17, 104–8; cell phones and, 114–15; in China, 105, 106; converting light into electricity, 104; in Germany, 58–59, 105; historical background, 104–5; in Japan, 43–44, 105; in poor and rural areas, 114–15; smart green grids and, 124–26; in South Korea, 105; in the U.S., 105–7, 130–31. *See also* Photovoltaic (PV) systems
Solar panels, 12, 67
Solarpark Hemau (Germany), 105
Solar-Powered Rooftops Plan (China), 69
Solar Valley City (China), 68–69
SOLON, 54
South Korea: G3IR in, 1–2, 7, 12, 46–48; solar power in, 105
Spain, 85
Steam, 97
Steam engine, 4
Steinfeld, Henning, 28
St. Mary's Cement plant (Ontario), 144–45
Storage technologies, 15, 16, 19–21, 51, 125. *See also* Intermittent technologies
Subsidies: in China, 68, 165–66; for oil companies, 130, 150–51
Sulfur hexafluoride, 35
Sundra Solar Corporation (China), 69
Supply-side economics, 55
Sustainability, in China, 66–67
Sustainable communities, 1, 15–16, 59–62, 77–95, 173–76; agile, 83–84; creating and implementing, 92–95; distributed energy generation, 59–62; G3IR in poor countries, 114–15; home/family behaviors and, 80, 84–86; hotels and resorts as, 90–92; infrastructure elements of, 81; Leadership in Energy and Environmental Design (LEED), 82; local/regional level components, 81–82; Melbourne, Australia, 86–89; origin of, 79; schools, colleges, universities, and other building clusters as, 89–90; sustainable development defined, 80; three basic concerns, 60. *See also* Commercial technologies and G3IR; Onsite power systems; Renewable energy; Smart green grids
Switchgrass, 143, 146
"Symbolic interaction," 50

Taliban, 150
Task-ambient lighting placement, 132
Tax breaks: for commercialization of new technologies, 147; for oil companies, 130, 150–51, 167, 176; for solar and wind energy systems, 131, 170
Technological innovations, 129. *See also* Commercial technologies and G3IR; Renewable energy; Smart green grids
Technology. *See* Commercial technologies and G3IR
Technology Entertainment and Design, 10
Telephone, invention of, 4
Texaco, 167
Thatcher, Margaret, 49, 149
Thin film, 105
Third Assessment Report (UN IPCC), 32
Three Gorges Dam, 110–12
Tidal energy, 18, 111–12
TOTO, 44–45
Toyota Prius, 8, 84, 93–94
Training, 51
Tramcar motors, 139–40
Transformative technologies, 132; commercializing, 146–48. *See also* Commercial technologies and G3IR; Renewable energy; Smart green grids
Transmission lines, smart green grids and, 126–28
Tsunami and earthquake in Japan (March 11, 2011), 11, 45–46
Tupi Field (Brazil), 9–10
2IR. *See* Second Industrial Revolution

UN FCCC. *See* United Nations Framework Convention on Climate Change
UN IPCC. *See* United Nations Intergovernmental Panel of Climate Change
United Nations: Environment Programme, Green Jobs: Towards Decent Work in a Sustainable Low-Carbon World, 163; Food and Agriculture Organization, *Livestock's Long Shadow—Environmental Issues and Options*, 28; Food and Agriculture Organization, Livestock Information and Policy Branch, 28; Framework Convention on Climate Change (UN FCCC), 31–32; Intergovernmental Panel of Climate Change (UN IPCC), 4, 31–32; Intergovernmental Panel of Climate Change (UN IPCC), *First Assessment Report*, 31–32; Intergovernmental Panel of Climate Change (UN IPCC), *Fourth Assessment Report*, 32–35; Intergovernmental Panel of Climate Change (UN IPCC), *Second Assessment Report*, 32; Intergovernmental Panel of

Climate Change (UN IPCC), *Third Assessment Report*, 32; investment in renewable energy worldwide (2009 estimates), 113–14; *Our Common Future*, 80; "World Population Prospects," 39–40
United States: coal use in, 97–98; energy use in (2011), 97; geothermal energy in, 109; Green Third Industrial Revolution (G3IR): the Asian approach and U.S. reaction, 48–52; moving toward G3IR, 7–22, 130, 167–78; post 9/11 in U.S., 1, 49, 150, 170–73; solar power in, 105–7; sustainable communities in, 86; wind farms in, 102, 130–31; wind turbine investment in, 164
United States Geological Survey (USGS), 9
University of California campuses, 145
Uranium, 11
USGS. *See* United States Geological Survey

Vanguard I, 104
Vectrix electric maxi-scooter, 140
Venture Business Research, 165
Venus study (Hansen), 29–30
Vestas, 56–57, 65, 154
Vroegh, Martin, 144

Walton, Alice, 3
Warming of the planet, 29, 34–35
Water energy, 15; hydroelectric power, 110–12; tidal power, 111–12
Water resources, 37–38
Watson, Thomas, 4
Watt, James, 4, 25–26, 170
Wave energy, 15
Wave power conversion, 18
Whitfield, William, 42
Wind farms, 17, 102–3
Windmills, 99–100
Wind power generation, 3, 15, 16–17, 99–104; in Chile, 102–3; in China, 70–72; in Denmark, 82; in England, 102; historical background, 99–102; smart green grids and, 124–26; in the U.S., 102, 130–31
Wind turbine manufacturing, 14, 56–57, 65; in China, 71; investment in U.S., 164. *See also* Vestas
Wireless technology, 120
Workforce green training, 64–65
"World Population Prospects" (UN), 39–40

Zero-emission engine, 137

About the Authors

WOODROW W. CLARK II, MA^3, PhD, is a long-time advocate for the environment and renewable energy as well as an internationally recognized author, lecturer, and advisor specializing in sustainable communities. Clark was one of the contributing scientists to the work of the UN IPCC (Intergovernmental Panel on Climate Change), which was awarded the Nobel Peace Prize in 2007.

GRANT COOKE, MJ, is an award-winning journalist, college administrator, and a pioneer businessman in the energy efficiency and energy renewable industries.